Biological Physics

D. C. S. WHITE

Department of Biology
University of York

LONDON

CHAPMAN AND HALL

First published 1974
by Chapman and Hall Ltd,
11 New Fetter Lane, London EC4P 4EE
© 1974 D.C.S. White
Set by E.W.C. Wilkins Ltd, London & Northampton
and printed in Great Britain by
Lowe & Brydone (Printers) Ltd,
Thetford, Norfolk

ISBN 0 412 12650 8 (cased edition)
ISBN 0 412 13600 7 (limp edition)

Distributed in the U.S.A.
by Halsted Press, a Division of John Wiley & Sons, Inc.,
New York

Biological Physics

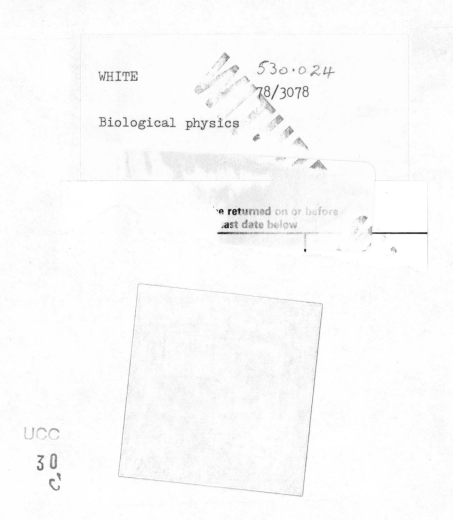

TO

Ailsa,
Kirsty and Stephen

Preface

This book is an introduction to the physical principles encountered by biologists, and is intended to be intelligible to those without advanced school physics. It differs from many texts with a similar title in being basically a biological book dealing with the necessary physics, rather than a physics textbook with the unnecessary chapters removed. As such, a number of topics not normally encountered in an elementary physics book are introduced; for example, the chapter on 'Deformation of solids' includes sections on rubber elasticity and viscoelasticity, the chapter on 'The motion of fluids' discusses a variety of phenomena in flight, and the chapter on 'Radioactivity' includes sections on the biological effects of radiation and indicates the way in which the reader can calculate the danger from radioactive samples. There is a chapter on energetics which takes as its starting point the concept of chemical potential energy, and proceeds with the thermodynamics required by the biochemist, the physiologist, the botanist and the environmental biologist in a unified way not always apparent from their separate treatment in specialized books. The chapters on electricity and magnetism introduce the concepts necessary for using electronic apparatus, pointing out the parameters that must be understood when using such apparatus as oscilloscopes and amplifiers, without fussing over detail of their design.

Other features of the book are the extensive sets of tables of values of the physical properties needed, and the questions at the end of most of the chapters, with fully detailed solutions. SI units are used throughout (and a foolproof method of converting units is given in Chapter 2).

D. White

Acknowledgements

This book originated as a set of papers provided to undergraduates at the University of York attending a lecture course on Physics for Biologists during their undergraduate Biology degree. The bulk of the book has been so used for three years, and during that time has had numerous improvements and additions. I am indebted to all those undergraduates who have provided a constant stimulus to me, and to all the numerous points and questions they have raised. I am likewise indebted to the staff here in York who have been pestered by me for ideas and examples. Above all I would like to thank Professor John Currey who has read most of the book and made innumerable suggestions. He has been the ideal critical reader. It also gives me much pleasure to thank Mr Geoffrey Pearce and Mrs Madeleine Donaldson who between them drew many of the diagrams and helped solve the problems, and Mrs Margaret Britton and Mrs Joan Chambers who typed much of the typescript.

Most of all however I would like to thank my wife who has kept our children happy whilst I have been writing the book.

Contents

Tables

UNITS AND DIMENSIONS

Variable	Name	Abbrev.	SI units Units	Dimension	Other units
Length	metre	m	m	l	inch (2·54 cm)
Mass	kilogram	kg	kg	m	pound (0·454 kg)
Time	second	s	s	t	hour
Area			m^2	l^2	
Volume			m^3	l^3	
Velocity			m/s	lt^{-1}	
Acceleration			m/s^2	lt^{-2}	
Frequency	hertz	Hz	s^{-1}	t^{-1}	
Density			kg/m^3	ml^{-3}	
Moment of inertia			kg/m^2	ml^{-2}	
Force	newton	N	$kg\ m/S^2$	mlt^{-2}	dyne (10^{-5} N)
Work, energy	joule	J	Nm	ml^2t^{-2}	erg (10^{-7}J), calorie (4·19J), eV (1·6 $\times$ 10^{-19}J),
Power	watt	W	J/s	ml^2t^{-3}	horsepower (746W)
Torque			Nm	ml^2t^{-2}	
Pressure			N/m^2	$ml^{-1}t^{-2}$	mmHg = torr (133 N/m^2) atmos (1·01 $\times$ 10^5 N/m^2)
Viscosity			Ns/m^2	$ml^{-1}t^{-1}$	Poise (10 Ns/m^2)
Surface tension			$N/m, J/m^2$	mt^{-2}	
Current	ampere	A	A		
Charge	coulomb	C	As		
E.M.F.	volt	V	W/A		
Resistance	ohm	Ω	V/A		
Capacitance	farad	F	As/V		
Electric field			V/m = N/C		
Magnetic induction			$N/(Am) = Weber/m^2$		Gauss
Radioactivity	curie	Ci	s^{-1}		1 Ci = 3·7 $\times$ 10^{10} d.p.s.
Light intensity	candela	cd	cd		

1 Mathematics

Many of you, reading this book, will have been taught a fair amount of mathematics, including calculus. You will have no trouble at all with the mathematics used in this book, which is kept at an elementary level. Some of you however will have done no calculus before, and may be mystified therefore when meeting some of the mathematical expressions used in the book. Take heart. For one thing such expressions do not occur too frequently, and furthermore when they do occur they are straightforward. The purpose of this chapter is to enable those of you who have not yet been taught any calculus to understand sufficient to be able to understand that in the book. The chapter is not intended to be a general introduction to calculus; for such a purpose you must read one of the many texts on mathematics such as those suggested in the Further Reading. Quite honestly, you would be well advised to do this anyway. This chapter is too short to be more than an introduction to the key ideas.

1.1 Finding slopes or gradients. 'The rate of change of . . .'

Let us suppose that you have just done an experiment of some kind, and that you have plotted the results from this experiment on a graph. To do this you will have made measurements of one variable, which you will have plotted on the y-axis, at different values of another variable, which you have plotted on the x-axis. For example, you may have measured the length of a muscle (y-axis) at different times (x-axis) after stimulating it to contract, or you may have measured the concentration (y-axis) of the product of a reaction after starting the reaction by the addition of enzyme, or maybe the number of animals in a given population (y-axis) at different times (x-axis). A hypothetical result is plotted in Fig. 1.1.

In order to be able to analyse the results of this experiment you may need to determine the *rate of change* of the variable on the y-axis; in the experiment on the muscle you might need to know the velocity of shortening of the muscle, in the experiment on the biochemical reaction you might require knowledge of the rate of the reaction (how many product molecules are reacting per second), and in the animal population the information needed could well be the rate at which the

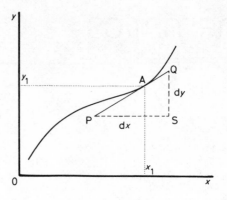

Fig. 1.1

population is changing. Suppose, in Fig. 1.1, you want to find the rate of change of the variable on the y-axis at the point A. The 'rate of change' of this variable is given by the slope of the curve at that point. The line PQ, drawn parallel to the curve at

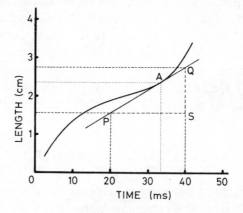

Fig. 1.2(A)

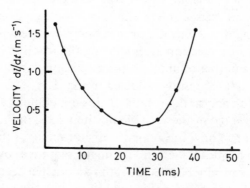

Fig. 1.2(B)

the point A, gives the value of the slope at this point. P and Q can be any points on this line; the slope of the line is then defined as

$$\text{Slope of line PQ} = \frac{QS}{SP} \tag{1.1}$$

The length QS is the difference in the values of y at Q and S, and we abbreviate 'difference in y' to dy. Likewise PS is the difference in the values of x at P and S which we abbreviate to dx. Thus we can say that the slope of the line is

$$\text{Slope} = \frac{\text{difference in } y}{\text{difference in } x} = \frac{dy}{dx} \tag{1.2}$$

pronounced 'dee y dee x'.

For example, Fig. 1.2A shows the same shaped curve as the results of an experiment in which muscle length (l) is plotted against time (t). We want to find out how fast the muscle is shortened at the point A. We draw the line PQ parallel to the curve at point A. At point Q the length of the muscle is 2·74 cm, and at point S it is 1·55 cm. Thus the 'difference in length' = dl = 2·74 − 1·55 = 1·19 cm. Likewise the 'difference in time' = dt = 40 − 20 = 20 ms. So the velocity at point A is

$$\text{velocity} = \text{slope} = \frac{dl}{dt} = \frac{11 \cdot 9}{20} = 0 \cdot 595 \text{ mm ms}^{-1}$$
$$= 0 \cdot 595 \text{ m s}^{-1}$$

Obviously we could, if we wanted, work out the slope of the curve at every point, and we could then plot a curve of this slope at the different times. This has been done in Fig. 1.2B by working out the slope for the points, and then drawing a smooth curve between them. This is then a curve of velocity against time.

In order to analyse the experiment further we might also need to know the *acceleration*. In order to work out the acceleration (which is the rate of change of velocity) we repeat the whole process, but using the curve of velocity against time. Thus, having drawn a line parallel to the velocity curve the acceleration is equal to the 'difference in velocity'/'difference in time' for two points on that line.

$$\text{Acceleration} = \frac{\text{difference in velocity}}{\text{difference in time}} = \frac{dv}{dt} \tag{1.3}$$

We can play around with the symbol 'd' meaning 'difference in', and substituting $v = dl/dt$ into this equation we get

$$\text{Acceleration} = \frac{dv}{dt} = \frac{d \cdot dl}{dt \cdot dt} = \frac{d^2 l}{dt^2} \tag{1.4}$$

which is pronounced 'dee two ell dee tee squared'.

1.1.1   Differentiation

For many purposes the above, graphical, process of finding slopes is the best to adopt. In fact with experimental data such as has been discussed so far you will almost certainly find slopes in this way. However, if the shape of the original curve can be expressed mathematically, then it is also possible to work out the values for the slopes mathematically. Often, furthermore, you will have a mathematical expression (such as a physical relationship between various parameters), and will want to determine an expression for the slopes. The process by which this is done is known as *differentiation*, which just means finding slopes or 'the rate of change of one variable with respect to another'.

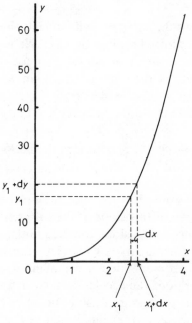

Fig. 1.3

For example, let us find an expression for the slope of the curve $y = x^3$, drawn in Fig. 1.3. For any value of x we know the value of y. Thus at $x = x_1$, $y_1 = x_1^3$. Now consider a point on the curve a *very* small distance (dx) away from x_1. (Notice that we have changed the meaning of dx slightly. Above we were simply meaning 'difference in x'; now we are meaning 'very small difference in x'). The value of y at this point will also have changed by a small amount dy. The co-ordinates of this new point are ($y_1 + dy$) and ($x_1 + dx$) and of course, since we know that for all points $y = x^3$,

$$(y_1 + dy) = (x_1 + dx)^3 \tag{1.6}$$

i.e. $\quad y_1 + dy \quad = x_1^3 + 3x_1^2\,dx + 3x_1 dx^2 + dx^3 \tag{1.7}$

but $\quad y_1 \qquad\quad = x_1^3 \tag{1.5}$

so $\qquad dy \quad = \qquad 3x_1^2\,dx + 3x_1 dx^2 + dx^3 \tag{1.8}$

Now comes the cunning bit. If we make dx small enough, then dx^2 and dx^3 are very much smaller even than dx. For example, if $dx = 0{\cdot}001$, then $dx^2 = 0{\cdot}000001$. For this reason we make dx very small indeed and consider that the terms with dx^2 and dx^3 are much less than that with dx, and we say that they can be ignored, leaving

$$dy \quad = \qquad 3x_1^2\,dx \tag{1.9}$$

The slope of the curve is given by dy/dx, and so dividing both sides of the above expression by dx we get

$$\frac{dy}{dx} = 3x^2 \tag{1.10}$$

This is then the expression we wanted for the slope of the curve, and $3x^2$ is known as the *differential* of x^3. For example, the slope of the curve at $x = 4$ (at which point $y = 64$) is $3 \times 4^2 = 48$.

This above procedure can be done for any mathematical expression that we like. Obviously the procedure would be tedious to go through every time we wanted to find the differential of an expression, and instead it is normal to look up the answer in tables. The differentials of a few expressions are given in Table 1.1, and this table includes all the forms you will need in this book. There are a few problems at the end of the chapter, and a full working of the solutions is given at the end of the book.

This is probably as far as you will need to understand differentiation in order to understand the book. However, your ability to differentiate will be considerably enhanced by a knowledge of three rules to be obeyed when meeting more complex expressions (these rules are also summarized in Table 1.1)

1. Differential of two terms multiplied together. If the two terms are u and v (which are both functions of x)

$$\frac{d(uv)}{dx} = u \cdot \frac{dv}{dx} + v \cdot \frac{du}{dx} \tag{1.11}$$

e.g. to find the differential of $x^3 \cdot \sin(kx)$. Let $u = x^3$, and $v = \sin(kx)$. Then $du/dx = 3x^2$ and $dv/dx = k \cdot \cos(kx)$

i.e $\quad \dfrac{d(x^3 \cdot \sin(kx))}{dx} = x^3 \cdot k \cdot \cos(kx) + 3x^2 \cdot \sin(kx) \tag{1.12}$

2. Differential of one term divided by another term.

$$\frac{d(u/v)}{dx} = \left(v \cdot \frac{du}{dx} - u \cdot \frac{dv}{dx}\right)\bigg/v^2 \tag{1.13}$$

e.g. to find the differential of $\cos(kx)/x^2$. Let $u = \cos(kx)$ and $v = x^2$. Then $du/dx = -k \cdot \sin(kx)$, $dv/dx = 2x$ and $v^2 = x^4$.

i.e.
$$\frac{d(\cos(kx)/x^2)}{dx} = \frac{x^2 \cdot (-k \cdot \sin(kx)) - \cos(kx) \cdot 2x}{x^4}$$

$$= \frac{-k \cdot \sin(kx)}{x^2} - \frac{2 \cdot \cos(kx)}{x^3} \tag{1.14}$$

3. Differential of the function of a function of x. If u is a function of x and $f(u)$ is a function of u, then

$$\frac{d(f(u))}{dx} = \frac{d(f(u))}{du} \cdot \frac{du}{dx} \tag{1.15}$$

e.g. to find the differential of $\sin(x^4)$. Let $f(u) = \sin(x^4)$ and $u = x^4$ i.e. $f(u) = \sin(u)$, with $u = x^4$. Then $d(f(u))/du = \cos(u)$ and $du/dx = 4x^3$. Thus

$$\frac{d(\sin(x^4))}{dx} = \cos(u) \cdot 4x^3 \quad \text{with} \quad u = x^4$$

$$= 4x^3 \cdot \cos(x^4) \tag{1.16}$$

Remember that in all these examples you have simply found the slope of the expression. If you need the slope at a particular value of x, then insert this value of x into the expression, and work out the value.

1.2 Maxima and minima

A graph will often have peaks and troughs (see Fig. 1.4). The peaks are known as maxima and the troughs are said to be minima. The slope of the graph at a peak or trough is zero (i.e. there is no change in y as x increases). The positions of peaks or troughs can thus be found by determining the expression for the first differential, and finding the points at which this is zero. This determines the points at which there are peaks or troughs, but it does not distinguish between them. To do this it is necessary to find the expression for the second differential, and determine its value at the positions of the maxima or minima. The slope of the dy/dx curve is d^2y/dx^2. The second differential will be positive for minima and negative for maxima. This can be seen from Fig. 1.4B which is a plot of the first differential of Fig. 1.4A (i.e. of the

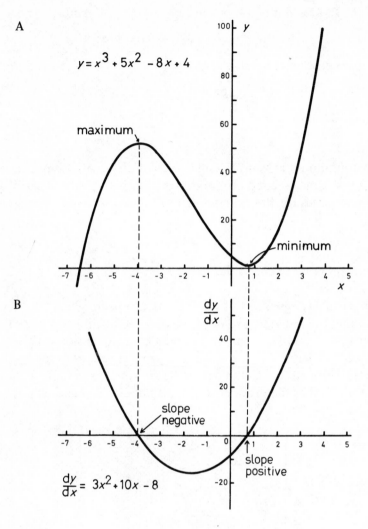

A

$$y = x^3 + 5x^2 - 8x + 4$$

maximum

minimum

B

$$\frac{dy}{dx}$$

slope negative

slope positive

$$\frac{dy}{dx} = 3x^2 + 10x - 8$$

Fig. 1.4

slope of the curve in Fig. 1.4A). The curve of dy/dx is zero at the values of x of the maximum and the minimum, has a negative slope at the value of x of the maximum and a positive slope at the value of the minimum.

Example. Find the maxima and minima of the expression

$$x^3 + 5x^2 - 8x + 4$$

first differential $\qquad 3x^2 + 10x - 8$

Find the values at which this is zero, by setting the expression equal to zero, and solving for x:

$$\text{Expression is zero when } 3x^2 + 10x - 8 = 0$$

$$(3x - 2)(x + 4) = 0$$

$$\text{i.e. when } x = +2/3 \text{ or } x = -4$$

second differential $\qquad\qquad 6x + 10$

$$\text{when } x = 2/3, \quad 6x + 10 = 14$$

$$x = -4, \quad 6x + 10 = -14$$

Thus, there is a minimum at $x = 2/3$ (since second differential is positive), and a maximum at $x = -4$ (since second differential is negative). You can easily check this by plotting out values for the expression, as has been done for this expression in Fig. 1.4.

1.3   Summation and areas under curves

There are a number of examples in this book of situations in which a total response is found by summing the individual responses from several contributory sources. A good example of this is found in Chapter 5 in which an expression for the work done in stretching a spring is derived. A spring has the property that the force it exerts (its tension) is directly proportional to the extension from its rest length, as shown in Fig. 1.5. Work is defined as the product of the distance moved against a force and

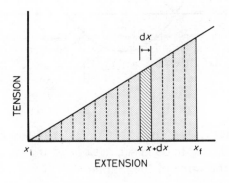

Fig. 1.5

that force. Consider the work done is stretching a spring which is already at an extension x by a small amount dx. Since the extra extension is small there is only a very small change in the force exerted by the spring and thus the work done for this extra extension (force times distance) is equal to the area of the hatched segment in Fig. 1.5. We can see therefore that the total work done in stretching a spring from rest length to an extension x_1 is the *sum* of the areas of such segments, equalling the total area under the curve (shown shaded).

The area of the hatched curve in Fig. 1.5 is $T_x \cdot dx$. The tension is directly proportional to the extension x; i.e. $T_x = kx$, in which k is a constant of proportionality known as the stiffness. The area of the hatched curve is thus $kx \cdot dx$, and the total work done in stretching the spring is:

$$\text{Total work} = \text{sum of } (kx \cdot dx) \text{ for all segments.} \qquad (1.17)$$

Another example (Chapter 6) is that in which an expression is derived for the rate of flow of fluid flowing down a tube, given a knowledge of the size of the tube and the pressure difference between its ends. Due to the viscosity of the fluid the velocity of flow varies at different positions; at the walls the velocity is zero, and in the centre it is maximal, as shown in Fig. 1.6. The volume of fluid per second flowing down the pipe in a segment of width dr at any radius r is equal to the product of the velocity of flow v_r at that radius and the cross-sectional area of the segment ($2\pi r \cdot dr$). Thus there are different rates of flow at different diameters, and in order to determine the total flow we have to *sum* these different contributions.

$$\text{i.e. Total Volume/second} = \text{sum of } (2\pi r v_r \cdot dr) \text{ for all segments}$$
$$= \text{sum of } 2\pi r(v_r \cdot dr) \text{ for all segments} \qquad (1.18)$$

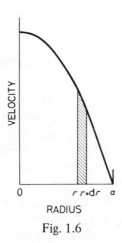

Fig. 1.6

Now, $v_r \cdot dr$ is the area of the hatched segment in Fig. 1.6. Thus, in principle, we can estimate the total volume by measuring the area of each of the segments in the graph of Fig. 1.6, multiply each by the relevant value of $2\pi r$ for that segment and sum the total.

1.3.1   Integration

In both of the examples in the previous section, the answer that we required was equal to the *sum* of a number of small contributions. The first example was the simpler, because the answer equalled the shaded area on the graph. The second example was more complex because the area of each small segment on the graph had to be multiplied by $(2\pi r)$, and so the final answer was not simply the area under the graph, although the final answer could have been worked out from the graph as described in the previous section.

As with the discussion on differentiation, if a mathematical expression can be provided for the shape of the graph, then a mathematical process, known as *integration* can be used to do the required summation for us without our having to resort to the graphs. Notice that in each of the two examples in the previous section the answer was written in the form

Answer = Sum of (expression describing one segment) for all segments (1.19)

These segments were delimited by two values of the variable along the x-axis; let us call these limits x_i and x_f. We can then say that:

Answer = Sum of (expression for one segment) for all segments
between the limits x_i and x_f.   (1.20)

Mathematicians use the *integral* sign $\int$ (which is simply an extended S for 'Sum of'), and they write the limits x_i and x_f below and above this sign as shown in Equation 1.21. The above equation, in its shorthand form, then becomes

$$\text{Answer} = \int_{x_i}^{x_f} (\text{expression for one segment}) \qquad (1.21)$$

When the limits x_i and x_f are specified as shown here the integral is known as a *definite integral*. We can also write the expression without the limits, in which case the integral is known as an *indefinite integral*, since until the limits are specified we cannot determine the actual *value* of the answer, although we can work out an expression giving us the general *form* of the answer.

In the same way as we were able to look up the differentials of an expression, so we can look up the integrals of an expression in tables, and a number of integrals, including all those that you will meet in this book, are given in Table 1.1. Notice that in this table the indefinite integrals are given. Thus for example the integral of x^2 is $\frac{1}{3}x^3 + C$. Here C is a constant, known as the *constant of integration*, whose value is unknown since we are just giving the general form of the answer. It is often omitted in tables of values of integrals.

In order to determine the value of the definite integral between two limits x_i and x_f we first of all look up the indefinite integral of the expression, and then subtract the value of the integral of the expression at x_i from that at x_f. For example, suppose that we wish to determine the value of the integral of x^2 between the values $x = 2$

TABLE 1.1

In this table, a and C are constants, x is the variable, and u and v are functions of x. ln is $\log_e$

Expression	$\dfrac{dv}{dx}$	$\int y \cdot dx$
a	0	$ax + C$
ax	a	$\frac{1}{2}ax^2 + C$
ax^n	nax^{n-1}	$\dfrac{1}{n+1}ax^{n+1} + C$
a/x	$-a/x^2$	$a \cdot \ln(x) + C$
e^{ax}	ae^{ax}	$\dfrac{1}{a}e^{ax} + C$
a^x	$a^x \ln(a)$	$a^x/\ln(a) + C$
$\ln(ax)$	$1/x$	$x \cdot \ln(ax) - x + C$
$\sin(ax)$	$a \cdot \cos(ax)$	$-\dfrac{1}{a} \cdot \cos(ax) + C$
$\cos(ax)$	$-a \cdot \sin(ax)$	$\dfrac{1}{a} \cdot \cos(ax) + C$
$u \cdot v$	$u\dfrac{dv}{dx} \times v \cdot \dfrac{du}{dx}$	
$\dfrac{u}{v}$	$\dfrac{u \cdot \dfrac{dv}{dx} - v \cdot \dfrac{du}{dx}}{v^2}$	
$f(u)$	$\dfrac{du}{dx} \cdot \dfrac{d}{du} f(u)$	
$u + v$	$\dfrac{du}{dx} + \dfrac{dv}{dx}$	$\int u \cdot dx + \int v \cdot dx$

and $x = 5$. The integral of x^2 is $\frac{1}{3}x^3 + C$. When $x = 5$, $\frac{1}{3}x^3 = \frac{125}{3}$. When $x = 2$, $\frac{1}{3}x^3 = \frac{8}{3}$.

$$\text{i.e. } \int x^2 \, dx = \tfrac{1}{3}x^3 + C \tag{1.22}$$

$$\int_2^5 x^2 = (\tfrac{125}{3} + C) - (\tfrac{8}{3} + C)$$

$$= \tfrac{125-8}{3} = \tfrac{117}{3} = 39 \tag{1.23}$$

Notice that in finding the value of the definite integral the value of C will always cancel.

In general then:

$$\int_{x_i}^{x_f} (\text{expression}) = (\text{integral of expression at } x_f)$$

$$- (\text{integral of expression at } x_i) \tag{1.24}$$

An important additional property of integrals is that the integral of the sum of two expressions is equal to the sum of the integrals of each expression by itself.

$$\int (\text{expression } 1 + \text{expression } 2) = \int (\text{expression } 1) + \int (\text{expression } 2) \quad (1.25)$$

Example

$$\int_0^a (Ax - x^3) \cdot dx = \int_0^a Ax \cdot dx - \int_0^a x^3 \cdot dx$$
$$= [(\tfrac{1}{2}Ax^2)]_0^a - [(\tfrac{1}{4}x^4)]_0^a$$
$$= \tfrac{1}{2}Aa^2 - \tfrac{1}{4}a^4$$

Here we have used square brackets with a superscript and a subscript $[\quad]_0^a$ to signify the value of the contents of the square bracket at the value of the superscript minus the contents at the value of the subscript. In this case, since we were dealing with definite integrals we knew that we did not need to include the constant C of the indefinite integral.

1.3.2   Integration as the inverse of differentiation

Integration is the inverse of differentiation. This means that if you differentiate the integral of an expression you end up with the expression itself.

$$\frac{d}{dx}\left\{ \int (\text{expression}) \, dx \right\} = \text{expression} \quad (1.26)$$

Example

$$\int x^3 \, dx = \tfrac{1}{4}x^4 + C$$

$$\frac{d}{dx}(\tfrac{1}{4}x^4 + C) = \frac{4x^3}{4} = x^3$$

A very useful check on possible errors in integration is to differentiate the integral and make sure that it gives the original expression.

1.4   Differential equations

As far as the contents of this book are concerned, differential equations are problems in integration. In a variety of the later chapters, a physical principle is expressed mathematically by an equation including a differential. For example, in Chapter 6, viscosity is expressed as the ratio of the shearing stress to the velocity gradient $\left(\dfrac{dv}{dz}\right)$ across a fluid. Thus, when determining the velocity of fluid flow along a pipe, at a particular radius r, the relationship is obtained (Equation 6.4)

$$\pi r^2 P = - 2\pi r l \eta \frac{dy}{dr} \quad (1.27)$$

It does not matter for the moment what all these symbols mean. The important point is that here is an equation relating the velocity v and the radius r. The equation includes the term $\dfrac{dv}{dr}$. We want to know the value of v as a function of r.

Another example in the book is concerned with diffusion (Chapter 9) in which the rate of diffusion is proportional to the gradient of the concentration $\dfrac{dC}{dx}$. This leads (Equation 9.21) to the following relationship between the concentration and radius inside a cylinder

$$\frac{dC}{dr} = \frac{Br}{2D} \tag{1.28}$$

Here, we have the differential term $\dfrac{dC}{dr}$; we want to find the value of C at any value of r.

Such equations as Equations 1.27 and 1.28 are known as *differential equations*, because they include a differential term. They happen in these two cases to be fairly simple, because the differential term is a first differential, and because it is possible to separate all the terms containing one of the variables in the differential on one side of the equation, and those containing the other variable on the other side of the equation. If we effect this operation the two equations above give

$$dv = -\frac{rP}{2l\eta} \cdot dr \tag{1.29}$$

$$dC = \frac{Br}{2D} \cdot dr \tag{1.30}$$

Each side of the equation can now be integrated between two values of the variables. The lower limit (corresponding to x_i in Equation 1.24) on the left hand side must be the value corresponding to the lower limit on the right hand side and likewise for the upper limits. Thus, in Equation 1.29, if we integrate the right hand side of the equation from the outer diameter of the pipe (at which $r = a$) to a radius z, then we must integrate the left hand side from the velocity at the outer diameter of the pipe to the velocity at $r = z$. Let the velocity at a radius z be v_z. The velocity at the outer diameter is zero, and thus:

$$\int_0^{v_z} dv = -\int_a^z \frac{rP}{2l\eta} \cdot dr \tag{1.31}$$

The integral on each side of the equation can be easily integrated (the integral of dv is v) giving

$$v_z - 0 = -\frac{P}{2l} \left(\tfrac{1}{2}z^2 - \tfrac{1}{2}a^2 \right)$$

$$v_z = \frac{P}{4l\eta} \left(a^2 - z^2 \right) \tag{1.32}$$

Equation 1.30 is exactly similar in form, but the limits of the integration are different. The answer is given in Chapter 9.

It will often be the case that the solution of the differential equation cannot be found as easily as for the two examples given here.

Good mathematicians work out the answers for themselves.

Good biologists go to good mathematicians.

1.5 Various formulae

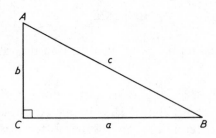

$$\sin A = a/c \qquad \mathrm{cosec}\, A = c/a$$

$$\cos A = b/c \qquad \sec A = c/b$$

$$\tan A = a/b \qquad \cot A = b/a \qquad \tan A = \frac{\sin A}{\cos A}$$

$$\sin^2 A + \cos^2 A = 1$$
$$\sin(A + B) = \sin A \cos B + \cos(A) \sin(B)$$
$$\sin(A - B) = \sin A \cos(B) - \cos(A) \sin(B)$$
$$\cos(A + B) = \cos(A) \cos(B) - \sin(A) \sin(B)$$
$$\cos(A - B) = \cos(A) \cos(B) + \sin(A) \sin(B)$$
$$\tan(A + B) = (\tan(A) + \tan(B))/(1 - \tan(A) \tan(B))$$
$$\tan(A - B) = (\tan(A) - \tan(B))/(1 + \tan(A) \tan(B))$$

$$\sin(A) + \sin(B) = 2\sin\tfrac{1}{2}(A + B)\cos\tfrac{1}{2}(A - B)$$
$$\sin(A) - \sin(B) = 2\cos\tfrac{1}{2}(A + B)\sin\tfrac{1}{2}(A - B)$$
$$\cos(A) + \cos(B) = 2\cos\tfrac{1}{2}(A + B)\cos\tfrac{1}{2}(A - B)$$
$$\cos(A) - \cos(B) = 2\sin\tfrac{1}{2}(A + B)\sin\tfrac{1}{2}(A - B)$$
$$\tan(A) + \tan(B) = \sin(A + B)/(\cos(A) \cos(B))$$
$$\tan(A) - \tan(B) = \sin(A - B)/(\cos(A) \cos(B))$$

$$\sin^2 A = \tfrac{1}{2}(1 - \cos(2A))$$
$$\cos^2 A = \tfrac{1}{2}(1 + \cos(2A))$$
$$\tan^2 A = (1 - \cos(2A))/(1 + \cos(2A))$$

$$\sinh(A) = \tfrac{1}{2}(e^A - e^{-A})$$
$$\cosh(A) = \tfrac{1}{2}(e^A + e^{-A})$$

$$\tanh (A) = \frac{\sinh (A)}{\cosh (A)} = \frac{e^A - e^{-A}}{e^A + e^{-A}}$$

For the general triangle

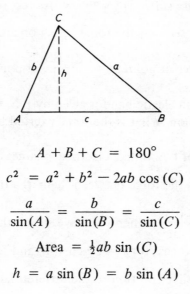

$$A + B + C = 180°$$

$$c^2 = a^2 + b^2 - 2ab \cos (C)$$

$$\frac{a}{\sin (A)} = \frac{b}{\sin (B)} = \frac{c}{\sin (C)}$$

$$\text{Area} = \tfrac{1}{2} ab \sin (C)$$

$$h = a \sin (B) = b \sin (A)$$

Circle of radius r. Area $= \pi r^2$

Circumference $= 2\pi r$

Ellipse of axes a and b. Area $= \pi ab$

Cone with radius of base r and height h. Volume $= \tfrac{1}{3} \pi r^2 h$

Sphere of radius r. Surface area $= 2\pi r^2$

Volume $= \tfrac{4}{3} \pi r^3$

Logarithms. If $y = 10^x$, then $x = \log_{10}^y$

If $y = e^x$, then $x = \log_e y = \ln y$

$$\log a + \log b = \log ab$$

$$\log a - \log b = \log \frac{a}{b}$$

Powers $a^n \cdot a^m = a^{(n+m)}$

$$e = 2\cdot71828$$

$$\pi = 3\cdot14159$$

$$\log_{10}^e = 2\cdot30259$$

1.6   Questions

1. Differentiate with respect to time t (a and b are constants)

   (i) $a \cdot \sin(bt)$ (ii) ae^{bt} (iii) $t \cdot \cos(bt)$ (iv) $e^{bt} \cdot \cos(at)$

   (v) $\dfrac{\cos(bt)}{t}$ (vi) $a \cdot \sin(bt^2)$ (vii) $a \cdot \cos(e^{bt^2})$

2. In chapter 4, page 34 we see that the force acting on an object is given by

$$\text{Force at point } a = \frac{d(\text{P.E.})}{dx} \text{ at point } a$$

a. The potential energy of an elastic band is given by P.E. $= \frac{1}{2}kx^2$ where k is the stiffness and x is the extension. What is the force exerted by the elastic band as a function of its extension.

b. The energy of a system of two unlike electrical charges e_1 and e_2 separated by a distance x is $\dfrac{e_1 \cdot e_2}{x}$. What is the force between them as a function of their separation.

3. If $y = a \cdot \sin(bt)$, what is $\dfrac{d^2 y}{dt^2}$? When you have solved this, write $\dfrac{d^2 y}{dt^2}$ in terms of y, eliminating t. Repeat this process for $y = c \cdot \cos(bt)$ and for $y = a \cdot \sin(bt) + c \cdot \cos(bt)$. Can you now suggest a possible solution to the differential equation $\dfrac{d^2 y}{dt^2} = -b^2 \cdot y$?

4. If $y = Ae^{at}$ what is $\dfrac{dy}{dt}$? Write an expression relating dy/dt to y. If in a population of animals the birth rate and death rate are both proportional to the population, with the birth rate exceeding the death rate, what is the relationship between population and time?

5. Solve the following

   (i) $\int a \cdot \sin(bt) \cdot dt$ (ii) $\int -\dfrac{a}{y} \cdot dy$ (iii) $\int b \cdot e^{at} dt$

   (iv) $\int b \cdot e^{ay} dy$ (v) $\int_0^{\pi/2b} a \cdot \sin(bt) \cdot dt$ (vi) $\int_0^{\pi/b} a \cdot \sin(bt) \cdot dt$

   (vii) $\int_0^1 x^2 \, dx$ (viii) $\int_3^4 x^2 \cdot dx$

6. Work our for yourself that the integrations of equations 6.4 (page 72) and 6.5 (page 73) are correct.

2 Units and Dimensions

Nearly all physical quantities have to be given units to make them meaningful. For example, when stating the distance between two points, it is no good giving the answer as 1·7; a unit must also be specified. The answer in this case might be 1·7 metres, written 1·7 m. Many different systems of units are in operation. Thus 1·7 m is the same length as 66·9 inches, 1·86 yards and 0·00106 miles. We can see that any physical quantity has a *dimension* (in the above example the dimension is length) and that to specify the magnitude of the quantity in this dimension the *unit* of measurement must also be specified, in this case metre or inch or foot or mile.

The dimension of any physical quantity can be defined in terms of six primary dimensions: length, mass, time, electric current, temperature and light intensity. Thus the dimension of area is (length)^2, abbreviated to l^2, and that of force is (mass × length/time^2) abbreviated to mlt^{-2}. Corresponding to each of the primary dimensions is a primary unit — an accepted standard unit measure of that dimension.

2.1 Primary and secondary units

The many different systems of units for physical quantities in current use can lead to considerable muddle. In an effort to produce a standard set of units acceptable to all disciplines an international and interdisciplinary commission has proposed a system known as SI (Système Internationale) units. These are now widely accepted, and many scientific bodies will only publish material using SI units. This book uses SI units throughout.

The six primary units are:

metre (m)   Unit of length
kilogram (kg)   Unit of mass
second (s)   Unit of time
ampere (A)   Unit of electric current
kelvin (K)   Unit of temperature
candela (cd)   Unit of light intensity

17

All other units are derived from these. Of the six, you will certainly have a good idea of the magnitude of one metre, one kilogram and one second. It is permissible to use the centigrade scale of temperature ($0°C = 273.16K$ = the temperature of triple point of water) with which you will be familiar. You may well not have any feel for the magnitude of one ampere or one candela. Do not worry.

The Frontispiece gives a list of other, secondary, units that you will use together with their dimensions. Also listed are units from some other scales to enable you to relate the new scale with what you may have been using before.

The fundamental definitions of the primary units, known as standards, are not likely to be of any practical use to you. The standards of length and time are defined in terms of atomic transitions, and the standard of mass is a platinum block kept (well locked) in Paris. Of much more importance from your point of view are the sources of the parameters that you will use experimentally. For accurate measurements of length a satisfactory laboratory standard is a cathetometer. This is a graduated metal ruler with a vernier scale and an attached travelling microscope. Cross hairs in the microscope eyepiece are used to align the microscope, whose position is then read on the calibrated rule. Accuracies of 0.01 mm are feasible. For mass a good set of balance weights will be accurate enough for most measurements. Clocks can readily be calibrated against broadcast time and used for time intervals greater than about one second. Short time intervals (less than one second) can best be standardized against the 50 Hz mains alternating voltage using an oscilloscope. Voltage can be accurately measured by comparison with a standard cell, and many voltmeters and other measuring devices contain a calibration cell of some kind. Resistance is measured by comparison with known resistors, using a Wheatstone bridge. Temperatures in the range $-10°C$ to $120°C$ can safely be measured to about $0.1°C$ accuracy with a good mercury-in-glass thermometer. Absolute accuracy to better than this is difficult, although estimates of temperature difference can be made to better than $0.0001°C$.

The secondary set of units can all be defined in terms of the primary set. Thus, the unit of force (the Newton) is defined as that force required to give a mass of one kilogram an acceleration of one metre/second^2. Its primary units are thus kg m/s^2, but this is abbreviated to N for newton. Units named after famous scientists are given capital letters. The precise definitions of these parameters will be given in the appropriate later chapters of the book.

2.2 'Cancelling' units

A physical quantity is written as a number followed by a unit. When, multiplying or dividing two or more physical quantities it is permissible to cancel units which appear both in the numerator and the denominator of the combined quantity, in order to determine the new unit.

For example, suppose we want to evaluate the force on a spherical molecule of radius α metres, moving with uniform velocity β metres/second through a fluid of viscosity μ Newton second/metre^2. Stoke's Law (see chapter 6, page 75) tells us that the force is

$$\text{Force} = 6\pi \text{ radius viscosity velocity}$$

$$= 6\pi \ \alpha m \ \mu \frac{Ns}{m^2} \ \beta \frac{m}{s}$$

$$= 6\pi \ \alpha \ \mu \ \beta \ \frac{m \ Ns \ m}{m^2 \ s}$$

$$= 6\pi \ \alpha \ \mu \ \beta \ N$$

Thus the units of the right hand side simplify to newtons, which is the correct unit of force.

2.3   Changing units

A similar kind of approach can be used if it is necessary to change from one set of units to another. Let us take as an example the process of converting 60 miles/hour into feet/second.

$$\text{Velocity} = 60 \frac{\text{mile}}{\text{hour}}$$

The procedure is first to write down the relationship between the individual units of the same dimension in the expression. In this case we have the conversion from miles to feet (dimensions of length), and from hours into seconds (dimension of time). We write:

$$1 \text{ mile} = 5280 \text{ feet} \qquad \text{and} \qquad 1 \text{ hour} = 3600 \text{ seconds}$$

If we divide one side of these two equations by the other we obtain expressions which have the numerical value of 1. Which side we make the numerator and which the denominator is dictated by the necessity to cancel units in the original expression. Thus:

$$1 = \frac{5280 \text{ feet}}{1 \text{ mile}} \qquad \text{and} \qquad \frac{1 \text{ hour}}{3600 \text{ seconds}} = 1$$

Since these expressions are both equal to unity we can multiply the original expression by both without changing its numerical value, giving:

$$\text{Velocity} = 60 \frac{\text{miles}}{\text{hour}} \frac{5280 \text{ feet}}{1 \text{ mile}} \frac{1 \text{ hour}}{3600 \text{ seconds}}$$

$$= \frac{60 \times 5280}{3600} \frac{\cancel{\text{miles}} \text{ feet } \cancel{\text{hour}}}{\cancel{\text{hour}} \cancel{\text{mile}} \text{ seconds}}$$

$$= 88 \frac{\text{feet}}{\text{second}}$$

2.4 Method of dimensions for determining and checking formulae

A quick check which can often show errors in a formulation is to compare the dimensions of each term in the expression. They must be the same. For example the total energy of a rigid body is given by the sum of its kinetic and potential energies.

$$\text{Total energy} = \tfrac{1}{2}mv^2 + mgh$$

in which the body of mass m is moving with velocity v at height h and g is the acceleration due to gravity. If this is correct, both terms on the right hand side of the expression should have the dimensions of the left hand side, that is, units of energy.

We will write m for the dimension of mass, l for length and t for time. The dimensions of $\tfrac{1}{2}mv^2$ are $m(l/t)^2 = ml^2t^{-2}$ which are the correct dimensions of energy. The dimensions of mgh are $mlt^{-2} \, l = ml^2 t^{-2}$ which are the same. This was a simple example. When you are working out more complex expressions for yourself you will find this check a very useful indication of possible errors.

Similar reasoning can enable the relationship between the various parameters in a formula to be determined. For example, suppose that we did not know Stoke's Law for the force required to keep a sphere moving through a fluid of a certain viscosity with a given velocity. Common sense would suggest that the force would depend upon the size of the sphere, the viscosity of the fluid and the velocity, but we would not be certain as to the relationship between these factors. We can write

$$\text{Force} = \text{const. radius}^x \text{ viscosity}^y \text{ velocity}^z$$

where x, y and z are unknown powers. The dimensions of the left hand side of the formula must equal the dimensions of the right hand side.

Thus

$$mlt^{-2} = (l)^x \, (ml^{-1}t^{-1})^y \, (l \, t^{-1})^z$$

This can only be true provided the dimensions of mass on the left-hand side equal those of mass on the right-hand side, and similarly for length and time.

i.e.

For mass	$m = m^y$	$\therefore y = 1$
length	$l = l^x l^{-y} l^z$	$\therefore x - y + z = 1$
time	$t^{-2} = t^{-y} t^{-z}$	$\therefore y + z = 2$

From these we find that $x = y = z = 1$, and thus the equation is

$$\text{Force} = \text{const. radius viscosity velocity}$$

This method cannot give values to the constants, and it is only valid provided all the variables affecting the equation can be obtained.

2.5 Multiples and sub-multiples of units

Often the value of a physical quantity will be must greater or much smaller than the value of one unit as defined in the Frontispiece. It is inconvenient for example to specify the distances between towns in metres since the numbers are all very large. Instead one uses a multiple of the basic unit, in this case kilometers where 'kilo' stands for 10^3. SI units use the scale of multiples and sub-multiples given in Table 2.1.

TABLE 2.1

Abbreviation	Prefix	Factor
T	tera	10^{12}
G	giga	10^9
M	mega*	10^6
k	kilo*	10^3
H	hecto	10^2
D	deca	10^1
d	deci	10^{-1}
c	centi	10^{-2}
m	milli*	10^{-3}
μ	micro*	10^{-6}
n	nano*	10^{-9}
p	pico*	10^{-12}
f	femto	10^{-15}
a	atto	10^{-18}

* These are the units you will meet most often.

2.6 Some useful physical constants

Table 2.2 gives a list of the values of some of the physical parameters you will need.

2.7 Questions

1. Given that there are 2·54 cm in one inch, convert 60 miles/hour into metres per second.

TABLE 2.2

1.	Atmospheric pressure (atm)	$1 \cdot 01 \times 10^5$	N/m^{-2}
2.	Acceleration due to gravity (g)	$9 \cdot 81$	m/s^{-2}
3.	Force to hold up a 1 kg mass	$9 \cdot 81$	N
4.	Gravitational constant (G)	$6 \cdot 67 \times 10^{-11}$	$Nm^2\ kg^{-2}$
5.	Gas constant (R)	$8 \cdot 32$	$JK^{-1}mole^{-1}$
6.	Avogadro's no. (N)	$6 \cdot 02 \times 10^{23}$	$mole^{-1}$
7.	Planck's constant (h)	$6 \cdot 63 \times 10^{-34}$	Js
8.	Boltzmann's constant (k)	$1 \cdot 38 \times 10^{-23}$	JK^{-1}
9.	Elementary charge (e)	1.60×10^{-19}	C
10.	Faraday (F)	$9 \cdot 65 \times 10^4$	$C\ mol^{-1}$
11.	Velocity of light (c)	3×10^8	m/s^{-1}
12.	Density of water	10^3	kg/m^{-3}
13.	Density of air	$1 \cdot 3$	kg/m^{-3}
14.	Absolute zero		$0°K = -273 \cdot 15°C$
15.	Electron-volt	$1 \cdot 60 \times 10^{-19}$	J
16.	Year	$3 \cdot 16 \times 10^7$	s
17.	Radius of earth	6400	km
18.	Radius of moon	1740	km
19.	Mass of earth	6×10^{24}	kg
20.	Mass of moon	$7 \cdot 4 \times 10^{22}$	kg
21.	Stefan's constant (σ)	$5 \cdot 69 \times 10^{-8}$	$W\,m^{-2}K^{-4}$
22.	Latent heat of vaporization of water	$2 \cdot 256 \times 10^6$	$J\,kg$
23.	Surface tension of water (20°C)	$7 \cdot 3 \times 10^{-2}$	$N\,m^{-1}$

2. The force on large objects moving through fluids in which the ratio of viscosity/density is small is determined only by the velocity of movement of the body relative to the fluid, the size of the object and the density of the fluid. Using the method of dimensions work out the formula relating the force to these three parameters.

3. Work out a conversion factor for converting the c.g.s. unit of force (the dyn = 1 gm cm/s^2) into newtons.

4. What are the dimensions of PV/RT where P is the pressure of gas in a container of volume V and temperature T and R is the gas constant (given in Table 2.2).

5. What is the dimension of Reynold's number, $Rn = \dfrac{\text{density} \times \text{length} \times \text{velocity}}{\text{viscosity}}$

3 Vectors

The Frontispiece lists a number of the variables that you will deal with in more detail in later chapters. It is convenient to distinguish these variables according to whether they are directional quantities or non-directional. Those variables that are directional (a *force* is an example — you push an object with both a certain magnitude and in a certain direction) are known as *vectors*; those described fully by their magnitude alone (such as *mass*) are *scalars*. Table 3.1 lists some parameters according to whether they are scalar or vector quantities.

TABLE 3.1

Scalar quantities	Vector quantities
Distance	Displacement
Speed	Velocity
Time	Acceleration
Mass	Force
Area	Electric Current
Volume	Electrostatic Field
Frequency	Electromagnetic Field
Viscosity	Momentum
Energy	Torque
Work	
Moment of Inertia	

Most of the classification is fairly obvious. Notice the distinction between distance and displacement, and between speed and velocity. Distance and speed are scalar quantities and just denote magnitudes; they are just numbers. Displacement and velocity are vectors and therefore must have direction specified also.

A vector can be represented diagrammatically by an arrow, or by a line which includes some representation of direction. The direction of the line is the direction of the vector, and the length of the line represents the magnitude of the vector. In written text vector quantities are also distinguished from scalar quantities, often by being

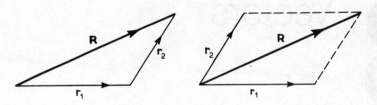

Fig. 3.1. Vector addition. $\mathbf{R} = \mathbf{r}_1 + \mathbf{r}_2$.

written in bold face. Thus $\mathbf{R}$ represents a vector, whereas R represents the magnitude
of the vector.

3.1 Vector addition

The sum of two scalar quantities is simply the sum of their magnitudes. Vector addition
is more complicated because the directionality must also be taken into account. It
can be achieved graphically by drawing the second vector with its base at the end of
the first vector. The vector sum $(\mathbf{R})$ is the vector obtained by drawing an arrow from
the base of the first vector $(\mathbf{r}_1)$ to the end of the second $(\mathbf{r}_2)$ as shown in Fig. 3.1. An
alternative method, giving the same answer, is to draw the two vectors originating
from the same point, and to complete the parallelogram. The vector sum is the
diagonal of the parallelogram from the base of the two vectors to the opposite corner,
as is also shown in Fig. 3.1.

  The easiest method of finding the vector sum of more than two vectors is an
extension of the first method described above; draw each vector in turn, starting the
second at the end of the first, the third at the end of the second and so on. The
resultant is found by joining the start of the first vector to the end of the last. This is
shown in Fig. 3.2.

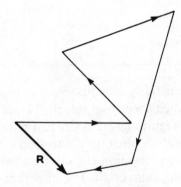

Fig. 3.2. Vector addition for more than 2 vectors.

3.2   Two examples to illustrate vector addition

(a) A locust flying through the air will have a certain velocity relative to the air. The air may have some velocity relative to the ground (wind). These velocities can both be represented by vectors. If we wish to determine the velocity of the locust relative to the ground this can easily be found by addition of the two vectors. In Fig. 3.1 above, r_1 might be the velocity of the locust relative to the air and r_2 that of the wind. The actual path over the ground taken by the locust will be **R**.

(b) In Fig. 3.3 is drawn the lower jaw of an animal, together with the vectors representing the forces applied by the muscles when it is eating (**T** denotes the temporalis and **M** the masseter muscle). The resultant of these two vectors (**R**) is determined in part B. If the jaw is in equilibrium the net force acting on it must be zero, i.e. the resultant force from the muscles must be balanced by an equal and opposite force, which in this case is the action of the upper jaw on the lower. This reaction is shown in part C.

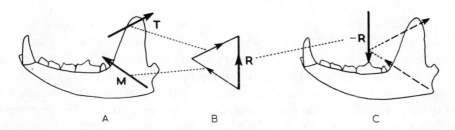

Fig. 3.3. A. Forces acting on the lower jaw of a possible mammal due to the action of the temporalis (**T**) and masseter (**M**) muscles. B. Resultant (**R**) of the vectors **T** and **M**. C. Reaction of the upper jaw on the lower. Vectors **T** and **M** are drawn dashed and their lines of action extended to show that **T**, **M** and $-$**R** all pass through a single point.

In this second example we determined the resultant vector by making use of the fact that the net force acting on the jaw was zero. From Newton's second law this means that the jaw is not accelerating; i.e. if it was stationary it will remain stationary even though great forces might be being applied to the food between the two jaws. Conversely, if after adding together the effects of several force vectors the result is zero (i.e. the end of the last vector just meets the start of the first) then we cay say that no net force is acting on that body.

Vector addition of forces is only valid provided the forces whose vectors are being summed all act through one point. In the example above Fig. 3.3 C shows that this condition was fulfilled in this case. If the forces do not all act through one point then the body will tend to rotate. Methods of dealing with this situation are dealt with in Chapter 4.

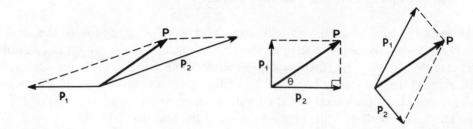

Fig. 3.4. Resolving a vector **P** into two equivalent vectors. When the two vectors are perpendicular they are said to be *component vectors* in their particular direction. They represent the net effect of a vector in that direction. If θ is the angle between a vector **P** and the direction along which the component is required then the magnitude of the component vector can be seen to be $P\cos\theta$. The component in the perpendicular direction will be $P\sin\theta$.

3.3   Resolving a vector into components

We have seen how to add two vectors. The method is illustrated in Fig. 3.1. Here vector **R** is equal to $\mathbf{r}_1 + \mathbf{r}_2$. It follows that any given vector is equivalent to an infinite number of pairs of vectors. Fig. 3.4 illustrates this idea. The vector **P** in each diagram is the same, and in each case it is equal to $\mathbf{p}_1 + \mathbf{p}_2$. The most generally useful way of resolving a vector into two equivalent vectors is that illustrated in the middle diagram. Here the two equivalent vectors are perpendicular to one another. The importance of this is that the effect of $\mathbf{p}_1$ has no effect in the direction of $\mathbf{p}_2$ and vice versa. This would not be true of the vectors of the other two diagrams.

3.4   Vector multiplication

(a) Scalar product

Suppose that we wish to evaluate the amount of work performed when a force **F** causes an object to move a displacement **d**. See page 30. Note that both the force and the displacement are vectors. (Work is often defined as the product of the force and the distance moved. This ignores the fact that the movement is in a particular direction, and that we should use a vector parameter to specify the movement.)

The work, which is a scalar parameter, is defined as the product of the displacement and the component of the force in the direction of the displacement. Previously we were only considering the force and the movement to be in the same direction, but this need not be the case. Suppose you pull a model railway truck along its rails (as in Fig. 3.5) but pulling at an angle, how much work is done? The truck will move only along the rails. The component of the force along the rail direction is $F\cos\theta$.

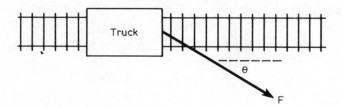

Fig. 3.5. Truck being pulled along rail track by force applied at an angle θ to the motion of the truck. Component of force in direction of movement is $F\cos\theta$.

If the distance moved by the truck along the rails is d then the amount of work done is $F\cos\theta \cdot d = F d \cos\theta$.

More generally, the *scalar product* of two vectors $\mathbf{R}$ and $\mathbf{S}$ is the product of their magnitudes times the cosine of the angle between them.

$$\text{Scalar Product } (\mathbf{R} \cdot \mathbf{S}) = R S \cos\theta$$

If the two vectors are in the same direction, then the angle between them is zero degrees. $\cos(0°) = 1$. Thus the scalar product of the two vectors in this case is the product of their magnitudes, as we expect.

(b) Vector product

It is also possible for the product of two vectors to result in a vector. In this case we say that we are taking the *vector product* of the two vectors (this is written with an inverted v between the vectors as $\mathbf{R}_\wedge\mathbf{S}$). An example of this is found in chapter 10 in which we see that the force on a charged particle moving in a magnetic field with velocity $\mathbf{v}$ is given by the vector product of the magnetic field $\mathbf{B}$ and the velocity $\mathbf{v}$. The magnitude of the resultant vector is the product of the magnitudes of the two vectors times the sine of the angle between them. The direction is perpendicular to the direction of the plane containing the two vectors.

If the two vectors are $\mathbf{R}$ and $\mathbf{S}$, then:

$$\text{Vector product } \times \text{ Magnitude} = \mathbf{R S} \sin\theta$$

$$\text{Direction Perpendicular to plane containing } \mathbf{R} \text{ and } \mathbf{S}$$

The reason for the $\sin\theta$ contribution to this rule, is that vector products are formed when the relevant interaction between the two vectors is between one vector and the component of the other perpendicular to the direction of the first. (Fig. 3.6)

3.5 Questions

1. At what angle to the wind must a locust fly in order to fly NE if the wind is due N

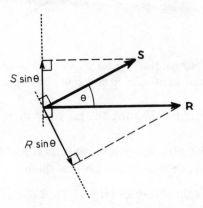

Fig. 3.6. Vector product $\mathbf{R}_\wedge\mathbf{S}$ has magnitude $R\,S\sin\theta = S\,R\sin\theta$ and direction coming vertically out of the paper.

with velocity 10 km/hour and the air speed of the locust is 15 km/hour?
2. A test-tube in a centrifuge is free to rotate in a horizontal plane. If the centrifugal acceleration is (a) 10 X gravitational acceleration (10 g), (b) 100 g (c) 10 000 g, what angle does the test tube make with the vertical?

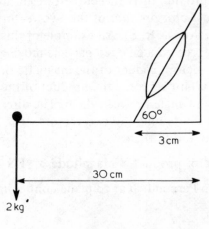

Fig. 3.7

3. A man is lifting a 2 kg weight as shown in Fig. 3.7. If the forearm weighs 0·5 kg, what tension in the muscle is required to maintain the position shown?

4 Mechanics

This section deals with the laws governing the movement of objects. It will cover the way that objects move as determined by the forces acting on them, and will discuss their corresponding energies. It is not intended to discuss the principles at great length, but to act as a convenient reference for definitions and laws and basic equations.

4.1 Mass

The mass of a body is a fundamental quantity representing the amount of matter in that body, and is constant for that body under all conditions (except if the velocity of that body approaches the speed of light — this is outside the scope of this book.)

4.2 Momentum

The product of the mass of a body and its velocity.

4.3 Conservation of momentum

If a system of perfectly elastic bodies is not acted upon by an external force, then the total momentum of that system remains constant.

4.4 Newton's laws of motion and force

1. Every body will continue in its state of rest, or of uniform motion in a straight line unless acted upon by a force.
2. The rate of change of momentum of an object is directly proportional to the imposed force acting on that object.

$$F = k\frac{d(mv)}{dt} \tag{4.1}$$

We shall only consider cases in which the mass of an object remains constant. Thus:

$$F = km\frac{dv}{dt} = kma \tag{4.2a}$$

in which a is the acceleration of the body. The *unit of force* in SI units is defined as that force required to give a mass of one kilogram and acceleration of 1 m/s^2, and therefore in SI units

$$F = ma \text{ newtons} \tag{4.2b}$$

3. If an object A exerts a force on an object B, then the object B exerts an equal and opposite force on the object A.

4.5 Pressure

A force applied uniformly over a given area is said to be exerting a pressure equal to the force per unit area.

$$\text{pressure} = \text{force/area}$$

'Pressure' is a word normally used in connection with the forces exerted by or on fluids (see Chapter 6). Atmospheric pressure is $1 \cdot 01 \times 10^5 \text{ N m}^{-2}$.

4.6 Work

Work is done when a force applied to a body causes that body to move. The amount of work is the product of the force F and the displacement S moved.

$$\text{Work} = FS \tag{4.3}$$

If the force is measured in newtons and the displacement in metres then the units of work obtained are joules. See also page 26.

4.7 Power

Power is the rate of doing work.

$$\text{Power} = \frac{dW}{dt} = \frac{d(FS)}{dt} \tag{4.4a}$$

If the force remains constant, then,

$$\text{Power} = F\frac{dS}{dt} = Fv \tag{4.4b}$$

The SI unit of power is the watt. One watt equals one joule/second. ($1\,\text{W} = 1\,\text{J/s}$)

4.8   Energy

Energy of a body is the ability of that body to do work, and is measured in terms of
the amount of work that can be obtained from that body. The units of energy are
therefore the same as those of work (joules). Energy can appear in many different
forms (e.g. kinetic or potential energy of a rigid body, light, heat, electrical energy,
energy within chemical bonds etc.). These can in principle be converted from one
to another, although the mechanisms for doing so are not always apparent.

4.9   Conservation of energy (First law of thermodynamics)

A fundamental axiom of all science is that energy is conserved (i.e. energy cannot be
lost or gained). There is no proof of this principle; its truth rests simply on the fact
that there is no known case in which its incorrectness is demonstrable.

4.10   Gravitational force

There are numerous different types of force, many of which you will meet in this
book. Some of them are strictly explainable in terms of more fundamental forces,
but are convenient concepts to use without such explanation every time they are
used. An example of this is the force exerted by a rubber (see Chapter 5). We define
a force here which is proportional to the extension of the rubber. This does not deal
as such with the way in which the force arises. Certain forces are fundamental in the
sense that they are basic laws of nature, and which cannot be broken down into
more fundamental types. One of these is the interaction between masses.
    It is found that two masses attract one another with a force given by:

$$F = G\frac{m_1 m_2}{r^2} \tag{4.5}$$

in which m_1 and m_2 are the two masses, separated by a distance r. G is a universal
constant, known as the gravitational constant, with a value $6{\cdot}67 \times 10^{-11}\ \text{N m}^2/\text{kg}^2$.
    No satisfactory explanation of why there should be this relationship is available.
It is accepted because there is no known condition in which it is not true.
    In order to use this formula it is also necessary to make use of the fact that a mass

occupying a large volume (e.g. the Earth) can be considered to be concentrated at the centre of gravity of that mass as far as using the above law is concerned.

Thus an object on the earths surface of mass m is attracted towards the centre of the earth with a force

$$F = G M m/r_e^2 = m (G M/r_e^2)$$ (4.6)

in which M is the mass of the earth and r_e is its radius. If we lift the object a few feet above the Earth's surface we have changed the value of r that we should use in the above formula by such a small amount that we can ignore the change. Thus objects on the earth's surface experience a virtually constant force. The value of $(G M/r_e^2)$ is known as the gravitational acceleration g. g has a value of about $9 \cdot 81$ m s^{-2}. Of course, objects such as satellites which leave the immediate vicinity of the earth no longer experience the same gravitational attraction. Men in satellites are often described as experiencing conditions of weightlessness. This does not mean that there is no gravitational attraction of the earth acting on them. In this situation their velocity is sufficiently great that the effect of this gravitational force is to produce an acceleration towards the centre of the earth which just maintains their orbit (see section 'Motion in a circle' on page 41).The feeling of weightlessness occurs because of the absence of constraints normally present on earth (such as the floor), resulting in the free acceleration. Thus, for example, there is no net force, for the man in the satellite, tending to pull him towards the floor of the satellite, because the satellite is accelerating away from him as fast as he is accelerating towards it.

4.11   Weight

The weight of a body is the gravitational force acting on the body. Thus the unit of weight is the same as the unit of force — the newton. Very appropriately, one apple weighs about one newton!

4.12   Potential energy

Potential energy is the energy that a body possesses by virtue of its position, and it arises under conditions in which there is a force acting on that body, trying to move it to a new position. It is correct to talk about the energy of a body subjected to different types of forces as being 'potential' energy. The force applying to macroscopic bodies is the gravitational attraction of the Earth, but we can also talk about the potential energy of, for example, ions subjected to electrical forces, nuclear particles subjected to nuclear forces and planets subjected to the gravitational forces of their 'Suns'. In each case the potential energy of the body at its particular position will depend upon the amount of work that would have to be done to move the body from this position to a reference position at which the potential energy is defined as being zero.

Let us consider the case of a mass m on earth. Provided that the heights involved are small compared with the radius of the earth then this has a constant gravitational force $m\,g$ acting upon it. The work (force × distance) that would have to be done on this body to lift it a vertical height h is thus $m\,g\,h$. If we define the position of zero potential energy as being say the floor, then the potential energy of a body at a position a height h above the floor is $m\,g\,h$.

This calculation was particularly easy because the force acting on the body in this situation is constant. For the other examples we used above this is not so. In these cases we have to use calculus to determine the amount of work done on the body to move it from one position to another. This will be illustrated by determining the potential energy of an ion with charge e a distance d from another ion of charge e_0. The force exerted by the one charge on the other, when the separation is x is

$$\text{Force} \; = \; k\,\frac{ee_0}{x^2} \tag{4.7}$$

in which k is a constant (see Chapter 7). The work done in moving one charge a small distance dx closer to the other charge is

$$\text{Work} \; = \; \text{force d}x \; = \; k\,\frac{ee_0}{x^2}\,\text{d}x \tag{4.8}$$

Notice that when the separation is very large the force between the charges is very small. The reference position, at which the potential energy is defined as being zero, is taken to be when the charges are separated by an infinite distance. The total work done (and thus the potential energy of the system) in moving one charge from an infinite distance d away is found by integrating the above expression for the work between the limits infinity (∞) and d.

$$\text{Potential energy} \; = \; \int_{\infty}^{d} k\,\frac{ee_0}{x^2}\,\text{d}x \; = \; -\,k\,\frac{ee_0}{d} \tag{4.9}$$

This is the potential energy of the system of two charges separated by a distance d. Such a system is storing energy because the charges are trying to move relative to one another — in the absence of the constraining force which must be holding them at that separation they would either move closer to one another (if they were unlike charges) or further apart (if they were like charges).

This was a particular example, evaluating the potential energy of a charged particle acted on by an electrostatic force. We can obtain a general relationship between a force and the corresponding potential energy. In order to determine the potential energy of an object at a position a due to a particular force we have to perform the calculation

$$P.E. \; = \; \int_{x_0}^{a} F(x)\,\text{d}x \tag{4.10}$$

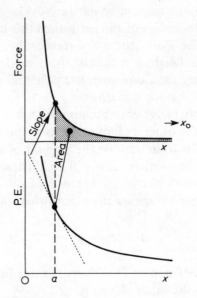

Fig. 4.1. Force and Potential Energy curves for an object subjected to electrostatic interaction (as between two charged objects). The force at any position is the slope of the P.E. curve at that position, as indicated in the figure, and the P.E. of an object at any position is the area under the force curve between that position and the point of zero potential energy, as is also indicated in the figure.

in which $F(x)$ is the magnitude of the force when the position of the object whose potential energy is being determined is x, and x_0 is the position at which the potential energy is defined as being zero. This enables us to determine the potential energy if we know the way in which the force varies with position. However, if, instead, we know the way in which the potential energy changes with position, then we can determine the force on the object. This is achieved by differentiating the above expression, giving

$$F(a) = \frac{\mathrm{d}(P.E.)}{\mathrm{d}x} \text{ at } x = a \qquad (4.11)$$

These two ideas can be expressed in graphical form (Fig. 4.1).

4.13   Translational and rotational movement

There are two different ways that an object can move

1. Translational movement, when all points in the body are moving in the same direction.

2. Rotation, when the object is spinning about an axis passing through the body. Different parts of the object are moving in different directions.

These two forms of motion can of course be superimposed, as for example in a wheel rolling along a road.

4.14 Kinetic energy (K.E.)

Kinetic energy is the energy that a body possesses because of its movement, and is defined as the amount of work that would have to be done to cause that body to become stationary. Since there are two forms of motion, there can be contributions to the total kinetic energy of the body from both, and we talk about the 'translational kinetic energy' or the 'rotational kinetic energy'. For the present we are going to restrict the discussion to translational movement only.

The value for the K.E. of an object of mass m, moving with a velocity v is most easily evaluated by applying a constant force F to the body and determining the distance over which this force must be applied to cause the body to stop. The amount of work done is then given by the product of this distance and the applied force.

The effect of an applied force F on a body of mass m is to give it an acceleration a given by $F = ma$. If the force slows the body down then we say that the acceleration is negative.

$$\text{i.e } F = ma = m\frac{dv}{dt}$$

We can now integrate this expression between our required limits.

$$\int m \, dv = \int F \, dt$$

Our limits are the initial velocity of the object (v) and zero velocity for the L.H.S. of the expression and the corresponding times for the R.H.S. Let us define both the time and the position of the body at the instant we apply our retarding force F as zero, and the unknown time and position when the body becomes stationary as T and x.

$$\int_v^0 m \, dv = \int_0^T F \, dt \tag{4.12}$$

Since both the mass and the applied force are constant they can be taken outside the integrals

$$m \int_v^0 dv = F \int_0^T dt \tag{4.13}$$

If we multiply both sides of the expression by the velocity of the body we get

$$m \int_v^0 v \, dv = F \int_0^T v \, dt = F \int_0^T \frac{dx}{dt} \, dt = F \int_0^x dx \tag{4.14}$$

where we have changed the variable (and the limits) of the R.H.S. of the expression.
Integrating both sides of the expression gives:

$$\tfrac{1}{2}mv^2 = F x \qquad\qquad (4.15)$$

The product $F x$ is the work done in stopping the body, and thus the kinetic energy
of the body moving with velocity v is $\tfrac{1}{2}mv^2$.

4.15   Rotational movement

So far all the discussion has concerned translational movement. All the objects
considered have in effect been thought of as single points. We shall now investigate
the rotation of objects, which means that we now think of them as having a finite
size, so that forces can act at different points on the same object. We shall only be
discussing rigid bodies in this section; this means that the object does not change its
shape or dimensions due to the action of imposed forces.

In principle the mechanics of rotational movement is very similar to that of
translational. Each of the different variables that we have defined has its rotational
counterpart. We measure movement in terms of angle, and it is normal to use radians
as the unit of angle rather than degrees. One radian is defined as the angle of the arc
(of a circle) whose circumference equals its radius. This gives a value of 1 radian of
about 57·3° Since the circumference of a complete circle is 2π (radius) it follows
that 2π radians equals 360°.

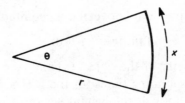

Fig. 4.2.

In general, if the angle subtended by an arc is θ, then the relationship between the
radius r and the length of the arc x is:

$$x = r\,\theta \qquad\qquad (4.16)$$

The rate of movement is measured in radians per second and is usually given the
symbol $\omega\ (= d\theta/dt)$. If we consider a brief instant of time then to a first approximation
the point at the end of the radius is moving in a straight line with velocity v (obtained
by differentiating equation

$$v = dx/dt = r\,d\theta/dt = r\omega \qquad\qquad (4.17)$$

Likewise, as we would expect, acceleration (a) is measured in terms of rate of change of angular velocity ($d\omega/dt = d^2\theta/dt^2$, often written $\omega' = \theta''$). We can write

$$a = dv/dt = d^2x/dt^2 = r\,d^2\theta/dt^2 = r\,d\omega/dt \qquad (4.18)$$

4.16 Torque (or moments)

The tendency of a force to produce rotational movement about a point is known as *torque*. Consider a child's see-saw (Fig. 4.3).

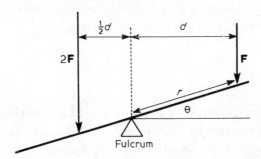

Fig. 4.3. A heavy child ($2F$) will balance a light one (F) if its distance ($d/2$) from the fulcrum is is half that (d) of the light child.

A heavy child can balance a light one provided that the heavy child sits nearer the fulcrum than the light one. Experimentally it is found that the ratio of the distances must be the inverse ratio of the masses to produce equilibrium. The torque is defined as the product of the force and the perpendicular distance between the force and the fulcrum (torque is a vector, defined as the vector product between the force and the vector joining the point of action of the force to the fulcrum.) In terms of Fig. 4.3.

$$\text{Torque } T = F\,d = F\,r\,\cos\theta \qquad (4.19)$$

Torque, also known as the *moment* of a force about a point, is the rotational equivalent of force.

If we consider rotation in a single plane only, then the rotation can be either clockwise or anti-clockwise. A torque tending to produce motion in a clockwise direction will have the opposite sign to one tending to produce motion in an anti-clockwise direction.

4.17 Couple

If two forces, equal in magnitude, but opposite in direction act on a rigid body

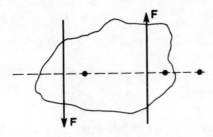

Fig. 4.4

through different points, then they exert a torque on that body, known as a *couple* which is a constant, independent of the axis of rotation.

4.18 Moment of inertia

This, in rotation, is the equivalent of mass. In translational motion each fraction of a mass requires that same fraction of force in order to be accelerated by a given amount. In the case of angular motion the effect of a given torque will depend upon the position of the mass. Consider a bar which is free to rotate about one end (see Fig. 4.5). Imagine this to be made up of thin slices, of mass m, along its length. Two of these are illustrated. A force F applied to that at r_1 will produce a linear acceleration $d^2 x/dt^2$, which is equivalent, using equation to an angular acceleration $r_1 d\omega/dt$ about the fulcrum, such that

$$F = m r_1 d\omega/dt \qquad\qquad (4.20)$$

The torque produced by the force is $F r_1$. That is

$$\text{Torque} = mr_1 d\omega/dt \; r_1 = mr_1^2 d\omega/dt \qquad\qquad (4.21)$$

This is the torque required to produce an angular acceleration $d\omega/dt$ of the slice at r_1. The total torque required to produce that angular acceleration in the whole bar is the sum of all the small torques required to accelerate the individual slices.

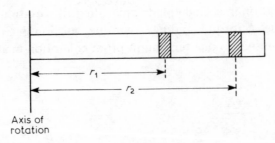

Axis of
rotation

Fig. 4.5

$$\text{Total torque} = \Sigma(mr^2) \; d\omega/dt \tag{4.22}$$

This is the angular motion equivalent of Newton's second law. The term $\Sigma(mr^2)$ is known as the moment of inertia I, and must be evaluated separately for every object which is rotated. Note that the moment of inertia of any particular object depends upon the point of rotation, and is different for different points of rotation. Table 4.1 gives some examples of the moments of inertia of common shapes of uniform density in terms of their dimensions and their mass M.

The equation of motion is

$$\text{Torque} = I \; d\omega/dt \tag{4.23}$$

TABLE 4.1

Shape	Body	Axis of rotation	I
	Rod	One end	$Ma^2/3$
	Rod	Centre	$Ma^2/12$
	Sheet	Centre. Perp. to sheet	$M(a^2 + b^2)/12$
	Circle Cylinder	Centre. Perp. to cylinder	$Mr^2/2$
	Circle	Diameter	$Mr^2/4$
	Sphere	Diameter	$2Mr^2/5$

4.19   Rotational momentum and energy

We can happily continue our analogy between translational and rotational terms. Angular momentum is the product of the moment of inertia and the angular velocity:

$$\text{Angular momentum} = I\,\omega \qquad (4.24)$$

The amount of work done against a torque is:

$$\text{Work} = \text{Torque} \times \text{Angle moved} \qquad (4.25)$$

The kinetic energy of a rotating body is:

$$\text{K.E.} = \tfrac{1}{2}I\omega^2 \qquad (4.26)$$

(When determining the total mechanical energy of a body this term must be included.) Table 4.2 lists various equivalences in the translational and rotational frames of reference.

TABLE 4.2

Translational	Symbol	Rotational	Symbol
Position	x	Angle	θ
Velocity	$v = dx/dt$	Angular velocity	$\omega = d\theta/dt$
Acceleration	$a = dv/dt$	Angular acceleration	$d\omega/dt$
Mass	M	Moment of inertia	I
Momentum	Mv	Angular momentum	$I\omega$
Force	F	Torque	$T\,(=F\,d)$
Work	Fx	Work	$T\theta$
Kinetic energy	$\tfrac{1}{2}Mv^2$	Kinetic energy	$\tfrac{1}{2}I\omega^2$

4.20   Equations of motion

It will perhaps be helpful to derive the equations of motion. These relate the initial (u) and final (v) velocities of a particle moving with constant acceleration (a) to the distance travelled (s) in that time (t).

Since the acceleration is constant we can write

$$\frac{d^2 x}{dt^2} = \frac{dv}{dt} = a$$

Integrate to give

$$\int_u^v dv = \int_0^t a\,dt$$

$$v - u = at \qquad (4.27a)$$

Integrate a second time to give (writing $v = dx/dt$ in (4.27a))

$$\int_0^s dx = \int_0^t u \, dt + \int_0^t at \, dt$$

$$s = ut + 1/2at^2 \qquad\qquad (4.27b)$$

Elimination of t from (4.27a) and (4.27b) gives

$$v^2 - u^2 = 2as \qquad\qquad (4.27c)$$

Elimination of a from (4.27a) and (4.27c) gives:

$$s = 1/2(u + v) \, t \qquad\qquad (4.27d)$$

4.21 Motion in a circle

A common form of motion of a body is motion round a circle (Fig. 4.6). Newton's first law tells us that a body will continue to travel in a straight line with a constant velocity unless acted on by a force. Thus in order to produce circular motion a force must be applied to the body to maintain the movement round the circle. It also follows that there will be a constant (sideways) acceleration on the body produced by this force (notice that the velocity and the acceleration, both of which are vector quantities are in different directions to one another).

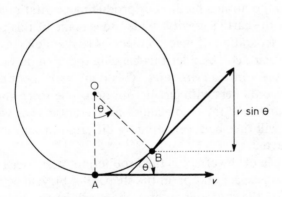

Fig. 4.6 Motion in a circle.

Consider a body of mass m at a particular instant of time which we will define as time zero, moving round a circle of radius r. Its angular velocity is ω. At time zero it will have a linear velocity v (equal, from equation (4.17), to $r\omega$) in a direction perpendicular to the line joining its position to the centre of the circle. A short interval, dt, of time later the body will have moved through an angle $d\theta$ round the circle, and it will have a new linear velocity, whose magnitude is still v, but whose direction has changed by $d\theta$. At time zero the component of the velocity in the direction AO of Fig. 4.6 is zero. At time dt the component is $v \sin(d\theta)$. For small angles the sine of the angle in radians is approximately equal to the angle $(\sin(d\theta) \doteq d\theta)$. Thus the change of velocity (dv) in time dt in the direction perpendicular to the original velocity is

$$dv = v \sin(d\theta) \doteq v \, d\theta \tag{4.28}$$

The acceleration in this direction is thus:

$$\text{acceleration } a = dv/dt = v \, d\theta/dt \tag{4.29}$$

Substituting for $v = r\omega$ (equation 4.17) and $\omega = d\theta/dt$ gives:

$$a = r\omega^2 \tag{4.30}$$

The force required to give this acceleration is:

$$F = ma = mr\omega^2 \tag{4.31}$$

in a direction perpendicular to the velocity at that instant. Thus the force is along the radius joining the object to the centre of the circle. This is termed *radial*. The force is known as the centripetal force.

Newton's third law shows us that if the body is acted on by a constant force F, then that body will exert a constant force $-F$ on whatever is applying the initial force. Thus the body which is being moved round the circle will exert a *centrifugal force* outwards, also of magnitude $mr\omega^2$.

Circular motion is the normal method of producing gravitational forces on a body different to those of the earth's gravity. Wide use is made of this principle. Thus a common method of separation of macromolecules (large biological molecules such as proteins and nucleic acids come under this heading for example) is to place the solution containing them into a *centrifuge*. This is an instrument designed for rotating test-tubes and other containers with circular motion. The force applied to the molecules will be $m\omega^2 r$, and thus molecules of different masses will experience different forces and will thus be accelerated by different magnitudes. This enables separation to be effected.

Another possible use of the effect is the production of artificial gravity in space stations. Imagine a large space station in the shape of a bicycle wheel, with the living quarters in the tube. If the station is rotated there will be a radial gravitational field set up. The dimensions of the space station need to be fairly large in order that the apparent gravity is sufficiently constant.

4.22   Friction and lubrication

When one object is moved across another it experiences a force tending to oppose the motion, known as friction. The laws of friction, which are empirical are:

1. The frictional force (f), parallel to the surface of contact between two objects, is proportional to the normal reaction (N) between the two surfaces. (The normal reaction is the magnitude of the component of the force perpendicular to the surface tending to keep the two objects in contact.)

$$\text{i.e. } f = \mu N$$

μ is the coefficient of friction. It is slightly larger whilst the objects are stationary than when they are moving, but is approximately independent of the velocity of movement.

2. The frictional force is independent of the area of contact between the two bodies.

Friction can be minimized by *lubrication*, which is essentially a separation of the two surfaces by the lubricant. The situation in which the lubricant is absorbed onto the surfaces of the two bodies, and in which there is little free lubricant between the surfaces, is known as 'boundary lubrication'. Long chain molecules give the lowest coefficients of friction in this situation. Lower coefficients of friction are obtained by 'fluid film' lubrication, in which the lubricant is at least many molecules thick. The coefficient of friction in synovial joints can be exceedingly low, often as little as 0·005.

4.23   Questions

1. Using values from Table 2.2 and equation 4.6 evaluate the gravitational acceleration at (a) the Earth's surface and (b) the Moon's surface.

2. What centrifugal force is exerted on a mass m kg at the Equator, due to the rotation of the earth?

3. The 'weight' of a body of mass 10 kg is measured with (a) a spring balance and (b) a normal beam balance supplied with a box of 'weights'. If the balances are calibrated in Oslo (Latitude 60°) what will be the difference in weight of the mass as measured by the two balances, at the Equator.

4. A uniform cylinder of radius 2 cm is allowed to roll down a plane 5 m long inclined at 30° to the horizontal, starting from rest. What is the final velocity of the cylinder?

5. A spaceship is in a steady circular orbit 2000 km above the surface of the earth. What is its velocity? What would happen if the velocity were suddenly increased by 10%? What is the minimum velocity required to leave the earth's gravitational attraction?

6. In order to simulate gravitational forces during a space flight it has been suggested that astronauts use a rotating spaceship. Work out suitable dimensions for this, together with the required frequency of rotation, given that it is essential that the acceleration at the astronaut's head must be no more than 1% different from that at his feet. What is the difference in direction of the gravitational force between the astronaut's semicircular canals?

7. A muscle is attached vertically to a mass of 100 gm supported on a table. Assume that in a twitch the muscle exerts a constant force of 5 N for 100 ms. Draw the timecourse of the change in position of the weight during the twitch. How much work was done by the muscle?

8. Prove that the moment of a couple about any point is a constant.

5 Deformation of Solids

Although life may have started in the aqueous phase, the evolution of larger and more complex organisms has required the manufacture of good structural materials. This chapter is concerned with the properties of these biologically interesting structural materials. As with so many branches of science, the precise study of a subject was initiated on the simplest items available, and this has meant that the biologist starts his study with a large background of similar work on non-biological material. Part of the fascination of the work to be looked into in this chapter has been the discovery of the number of times that nature has developed a more perfect form of a material that man has tried to manufacture artificially. Thus resilin is virtually the 'perfect' rubber and bone is an excellent two-phase material, which has its artificial courterparts in fibreglass and carbon-fibre reinforcement.

The study of the deformation of solids can ask questions at different levels. The first is simply to ask 'what happens if I try and extend, bend or twist this piece of material? How much force must I apply?' The experimental procedures can get quite complex as experiments to cover all eventualities are performed. A more fundamental question is to ask 'why is *this* amount of force needed to produce *this* deformation? What structures are responsible, and why?' We shall cover both approaches in this chapter.

Let us start by looking at some of the problems that organisms encounter in building structural materials. Substances with very different properties are needed for different functions. Thus many skeletal materials need to be fairly stiff in order that there is not too much deformation produced either by the weight of the body or by the action of muscles. Many animal phyla have members making use of comparitively rigid materials, from the crystalline materials used by protozoans and sponges to the cuticular structures of arthropods and the bones of vertebrates. Other animals require much less stiff skeletal materials, because their bodies change their shapes by large degrees, especially in locomotion. The hydroid coelenterates can go through incredible volume changes (especially when feeding), some nemertine worms can happily be extended to ten times their shortest length and the general hydrostatic skeleton, whilst not capable of such extreme length changes nonetheless undergoes appreciable deformations. Materials capable of coping with such effects are needed.

Not only must materials hold an animal together, they must also be made to withstand breaking when subjected to strains in all directions; exoskeletons need to avoid puncture by comparatively sharp objects and some animals need to be able to store energy in elastic structures (for example in flight), to name but three of the other functions.

Within any one animal a wide range of materials is needed. Think of the different mechanical properties required by man in his different structures – the rigidity and strength of bones, the larger extensibility and ability to be folded in all directions of skin, the elasticity of the body wall of the lungs, for example.

Obviously, therefore, a wide range of structural properties is required, and found, in the animal kingdom. The first thing that we must do in order to put some semblance of order in our understanding of these properties is to provide some precisely defined and experimentally measurable parameters.

5.1   Different ways of deforming solids

Basically what we are wanting to do is to define a relationship between the change in the dimensions of a material and the force required to cause that change. Part of the complexity of the subject arises because solids can be deformed in different ways. Fundamentally there are just three types of deformation; others can be considered as composite forms of these three. These three deformations (illustrated in Fig. 5.1) are

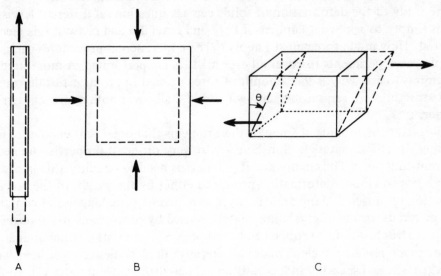

Fig. 5.1 The three basic deformations of a solid. A: Longitudinal extension; B: Volume change; C: Shear. Notice that longitudinal extension is accompanied by transverse contraction. The forces that must be applied to the specimens are shown.

longitudinal (one dimensional) length change, volume changes and shear. The forces required to give these deformations are also illustrated and are: for longitudinal length, equal and opposite forces applied at either end of the specimen and with the same line of action; for volume changes, a uniform force applied over the whole surface area (a change in pressure); for shear, a torque caused by equal and oppositely directed forces applied to opposite faces of the specimen (the lines of action of these forces will be separated by the width of the specimen).

Two particular deformations are also worth mentioning at this stage (Fig. 5.2). The first is torsion in a cylinder of uniform cross section (usually circular). This is a common way of producing shear in a specimen. Provided the twist caused by the torque is small then shear is the only deformation that need be considered. The other common deformation is bending, as of a long beam or of, say, a bone to which a sideways force is applied. Although at first sight it might appear that this is simply shear, this is not the case, and although there obviously is a shear force applied in this situation, provided the length of the specimen is large compared with its width, the major deformation is longitudinal. The top half of the beam will be extended, and the bottom half compressed, with a *neutral plane* of no extension in between. The extended top half will tend to shorten, and so produce a net force which in the diagram is towards the left; the bottom half will tend to lengthen and produce a net force to the right. These two forces produce a torque (or bending moment) which is in the opposite sense and equal in magnitude to the torque provided by the load (**P**) which is bending the beam. The extension or compression in any individual section is simply a longitudinal length change.

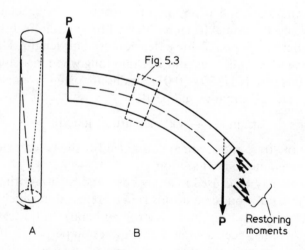

Fig. 5.2 A. Torsion in a circular rod. This is a method of producing shear in a specimen. B. Bending of a beam due to a load **P** applied at one end. The bending moments in the beam maintaining the equilibrium are also shown.

5.2   Tension

If a material is acted upon by forces tending to extend it as in Fig. 5.1A, then internal forces will be set up within the material resisting the applied external forces. These internal forces are known as the *tension* in the material, and have the same units as force (newtons).

5.3   Stress and strain

In order to obtain a given deformation of any of the three types we have to apply a given load. The magnitude of deformation and load will be dependent upon the size and shape of the specimen used for the test. Consider the longitudinal extension of a material. The thicker this material the more force we shall have to apply to produce a given extension. In fact the amount of force required to produce this given deformation will be proportional to the area of cross section of the material. We can see that this is sensible as follows. Suppose a force F produces an extension x in a single rod of material. If we have two equal rods side by side, whose ends are connected together, than we shall need $2F$ to produce the extension x. It does not matter whether the two rods are separate, or whether they are combined into one, the same force will be needed. Because of this it is standard practice to normalize the load by talking about the load per unit area of cross section. This is known as the applied *stress*.

$$(\sigma)\ \text{stress} \ = \ \text{applied load/unit area.} \tag{5.1}$$

Likewise the deformation will depend upon the length of the specimen. If we double the length, then we double the extension obtained for a given load. We normalize this parameter by defining *strain* as the extension per unit length. Thus if our specimen is 100 mm long when unloaded and 101 mm long when loaded (extension 1 mm) then the strain in this case is $1/100 = 0.01$. Strains are often given as percentage values. In this example the strain would be 1%.

$$\text{strain } (\epsilon) \ = \ \text{extension/unit length.} \tag{5.2}$$

Similar definitions of stress and strain are obtained for the volume deformation and for shear, and are given in the next section.

Notice that strains may be caused in a specimen in directions other than that of the applied load. The most obvious case of this is in a preparation which is loaded in one direction only (as, for example, pulling on a rubber band). In this case the specimen becomes thinner; there has been a strain in the direction perpendicular to the applied load. There will be no stress in this direction provided the preparation is in equilibrium. Thus care must be taken to distinguish between *applied* stresses and strains, and those resulting as a consequence.

5.4   Modulus of elasticity

For each of the three types of deformation we define a modulus of elasticity as:

$$\text{modulus of elasticity} = \text{stress/strain} \qquad (5.3)$$

where the stress and strain are the appropriate forms for that type of deformation.

TABLE 5.1

Approximate Values of Young's Modulus

Material	Y (N/m^2)
Steel	2×10^{11}
Wood	1×10^{11}
Rubber	3×10^{6}
Resilin	1×10^{6}
Bone	1×10^{10}
Spider's thread	3×10^{9}
Catgut	3×10^{9}
Silk	6×10^{9}

1. For longitudinal deformations the elastic modulus is known as *Young's modulus* and is generally given the letter Y or E to designate it.

$$Y = \text{applied load per unit area/extension per unit length.} \qquad (5.4)$$

In general when a material is loaded with a force applied in just one direction there are dimensional changes not only in that direction but also in the perpendicular directions. Thus when extending a bar by hanging a weight on it the bar will lengthen, and it will also become thinner. This effect can easily be seen by extending a rubber band. The area to be used in applying the above formula is that pertaining after the load is applied.

2. The elasticity obtained with volume deformations is known as the *bulk modulus*, and generally denoted by K.

$$K = \text{applied pressure/change in volume per unit volume.} \qquad (5.5)$$

3. The modulus in shear is known as the *rigidity modulus* (n) or (γ)

$$n = \text{tangential force per unit area/angular deformation.} \qquad (5.6)$$

Note that the area used in deriving the stress is the area of the surface to which the tangential force (or torque) is applied. Shear strain can be defined as the relative displacement of two planes whose separation is unity. If the angular deformation is θ then shear strain is strictly equal to $\tan(\theta)$. For small angles (less than about $10°$), $\tan(\theta)$ is approximately equal to θ, provided θ is measured in radians and hence its use in the definition of the rigidity modulus.

For the rest of this chapter we shall only discuss longitudinal extensions.

5.5   Measurement of Young's modulus by bending a beam

The importance of bending is that it is often used experimentally to obtain values
for Young's modulus. The reason for this is that for rigid materials the extensions
that can be applied are extremely small when the material is extended longitudinally,
and this makes accurate measurements difficult (and costly). Making the assumption
that the Young's modulus of the material is the same when the material is compressed
as when it is extended, the value for the modulus can be obtained by bending a beam
in which case easily measurable deflections are obtained. The assumption that must
be made is one of the drawbacks of this method. It is almost certainly not true in
many cases.

We shall derive the relationship between the bending of a beam and the Young's
modulus Y. Consider a short length of the beam of Fig. 5.2 as illustrated in Fig. 5.3.
The dashed line is the neutral plane which is unstrained. The radius of curvature of the
beam R is the radius of curvature of this plane, which subtends an angle θ as illustrated.

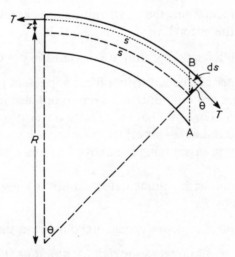

Fig. 5.3 Bending of a beam. A short section of a bent beam, subtending an angle θ is illustrated.
s is the length of the arc of the neutral plane.

The dotted line represents a plane a distance z above the neutral plane. The extension
of this plane is found by drawing the line AB parallel to the other side of the beam.
The strain in the plane is then ds/s. If the cross sectional area of the plane at z (as
seen from one of the cut surfaces) is α, then the tension exerted by this plane is given

(using the definition of Young's modulus) by

$$T/\alpha = Y\,\mathrm{d}s/s \qquad (5.7)$$

Since $s = R\,\theta$ and $\mathrm{d}s = z\,\theta$ (see Fig. 4.2 for explanation) it follows that

$$\mathrm{d}s/s = z/R \qquad (5.8)$$

Therefore

$$T/\alpha = Y\,z/R$$

$$T = \frac{Y}{R}\,\alpha z \qquad (5.9)$$

This tension exerts a moment $T \cdot z$ about the neutral plane

$$\text{Moment exerted by plane at } z = Tz = \frac{Y}{R}\,\alpha z^2 \qquad (5.10)$$

The total moment exerted by all the planes is found by summing this expression for all values of z

$$\text{Total moment} = \frac{Y}{R}\ \alpha z^2 = Y I_g/R \qquad (5.11)$$

$\Sigma \alpha z^2$ is known as the geometrical moment of inertia of the beam, I_g.

For a solid rectangle and a hollow cylinder the values of I_g are given in Table 5.2. NP is the neutral plane.

TABLE 5.2

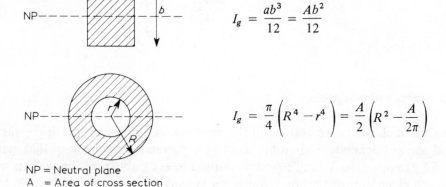

$$I_g = \frac{ab^3}{12} = \frac{Ab^2}{12}$$

$$I_g = \frac{\pi}{4}\left(R^4 - r^4\right) = \frac{A}{2}\left(R^2 - \frac{A}{2\pi}\right)$$

NP = Neutral plane
A = Area of cross section

The value $Y I_g$ is a measure of the resistance of the beam to bending and is known as the *flexural rigidity*. It is the bending moment required to produce unit radius of curvature.

Since the strain at different levels in the beam is different it follows that the beam will break first at either the top or the bottom surface, where there is maximal extension or compression. The *strength* of the beam to bending is proportional to I_g/R_{max} where R_{max} is the maximal distance of any part of the beam from the neutral plane.

The total restoring moment due to the bending of the beam will exactly balance the bending moment in the beam due to the applied load. This bending moment will vary down the length of the beam (it will be equal to the weight of the applied load times the horizontal distance between the point of application of the load and the point on the beam being considered), and thus the radius of curvature will vary down the length of the beam. It is usual to measure the deflection of the end of the beam experimentally; the relationship between this deflection and the bending of the beam at the different points along its length can be determined theoretically, and thus the Young's modulus determined.

Provided the length of the beam (l) is much greater than its width, then the deflection caused by a load P as in Fig. 5.2B is given by:

$$\text{deflection} = \frac{Pl^3}{3YI_g} \tag{5.12}$$

5.6   Buckling

If a long piece of thin material is compressed there will be a tendency for it to buckle rather than be properly compressed, due to any slight irregularity causing a bending in the material. The load at which the material will collapse is given by

$$P = \frac{\pi^2 YI_g}{l^2} \tag{5.13}$$

where l is its length.

5.7   Stiffness and compliance

The moduli of elasticity we defined in the previous section are properties of the material and not dependent upon the size of the particular specimen of that material. Given, for example, the Young's modulus of the insect 'rubber' resilin we can work out the actual amount of tension required to extend any given piece of resilin by a given length. We need to measure its length and its cross-sectional area; we then have sufficient information to evaluate the tension–length relationship. This can be done even if the specimen is not cylindrical, though the calculations will then become more difficult. Notice this distinction between the stress and strain on the one hand, and the tension and length on the other. The first gives the generalized terms which apply to all shapes and sizes, the second the particular terms to be used for an individual specimen. The ratio of stress to strain we termed a modulus of elasticity. The ratio of tension to length we term

$$\text{stiffness} = \text{tension change/length change} \qquad (5.14)$$

The inverse of stiffness is known as compliance.

$$\text{compliance} = \text{length change/tension change} \qquad (5.15)$$

5.8 Stress—strain curves

Another factor we have so far been rather imprecise about in this chapter has been the amount of stress or strain that we have been applying to our specimens. Thus, is the modulus of elasticity of a material a constant under any conditions we like to name? Does it, in particular, matter how *much* stress we apply, or how *fast* we apply it? The answer to both these factors is that it matters a great deal, and we must invesigate them a little more thoroughly.

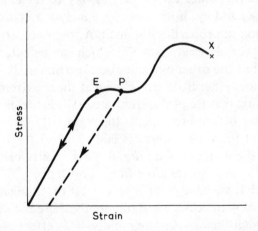

Fig. 5.4 Generalised stress—strain curve. E denotes the elastic limit, below which the curve is reversible. Beyond this, as at P, a reduction in length results in a new curve, shown dashed. The material finally breaks at X.

Suppose that we take a specimen of a material, and stretch it very slowly, at a constant rate, and measure the tension in the specimen continuously. What kind of curve do we get? Fig. 5.4 shows a generalized form of curve. In general there is a region in which the stress is directly proportional to the strain. If we double the strain on the specimen in this region we double the stress. This is said to be the region of *linearity*. If we stop stretching at any point up this curve and release the specimen slowly we go back down the same curve. At some point, labelled E in the figure, a limit is reached, beyond which the curve is no longer linear. This limit is known as the *elastic limit*, and the region below this point is known as the *elastic region* of the curve. It is in this region that the modulus of elasticity is obtained, and usually, with biological materials, this will be the region within which the animal uses the material.

Beyond the elastic limit irreversible things start happening. For one thing, if the specimen is released beyond this point, say at P then the tension falls too rapidly, and reaches zero with the specimen extended from its original starting point. There has been what is termed *plastic flow*. An exaggerated case of this is seen with plasticine or putty. Pull this and a new shape is obtained. This is not a liquid flow, because the shape obtained remains. It is due to a rearrangement of the molecules in the material.

If the specimen is extended further then there eventually comes a point (X) at which it breaks. The stress at this point is known as the *breaking stress* and the strain as the *breaking strain*. The breaking stress is also known as the *tensile strength* or *ultimate strength* of the material, or sometimes just the *strength*.

Fig. 5.5 illustrates some stress/strain curves of a variety of biological materials. At the top is a composite diagram giving the curves on a single diagram. This is obviously rather unsatisfactory since the properties vary to such an extent. The remaining parts show the curves separately. Both parts were included in order to emphasize the extremely wide variation in properties obtained. At the one extreme there are materials that produce very large stresses and which can be extended very little before they break, and at the other extreme there are materials which can be stretched a great deal, but produce rather little stress, even at these extreme strains.

Another distinguishing feature of different materials is the extent to which they will undergo plastic flow before breaking. If the region of plastic flow is very small, or virtually non-existent then the material is said to be *brittle*. If the region of plastic flow is extensive then the material is *ductile*. Bone is a fairly brittle material. Most metals are rather ductile. So also are most fibrous proteins.

One simplification that we have made is to suggest that the elastic region is always linear. This is by no means the case. Very frequently there is a curve which starts with a rather small slope which then rises more rapidly. This effect is reversible, in the sense that if the length of the specimen is reduced the same curve is obtained. In this case the modulus of elasticity quoted in tables is for the maximum elastic modulus before plastic flow sets in.

5.9   Dynamic effects

On page 53 we asked 'does the modulus of elasticity depend upon the rate at which the measurements are made?' This section deals briefly with this.

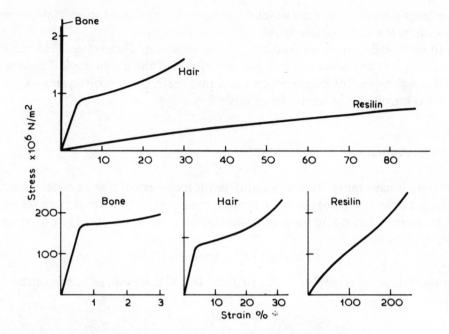

Fig. 5.5 Stress–strain curve of bone, keratin (from hair) and the insect rubber, resilin. In the top figure the three curves are all drawn on the same scale. That for bone is so close to the axis that it is not visible. In the lower curves the scales are changed and each curve is drawn separately. Notice the approximately tenfold change in the strain curves going from bone to keratin and from keratin to resilin.

These are two phenomena associated with performing the experiments more rapidly. The first is that the elastic modulus increases as the speed of measurement increases. That is, the more rapidly the measurements are made, the greater the tension produced for a given length change. The second is the effect known as *hysteresis*. We have suggested in the previous sections that in the elastic region of the material the same tension–length curve is obtained for increasing length as for an ensuing decrease in length. This is usually only the case provided the experiment is done extremely slowly. Much more usually the curve obtained as the length is decreased falls below that obtained as the length is increased, although the same length for zero tension will often still be reached. In other words the situation is rather more complex.

What makes the subject less tractable still is that the kinds of experiment usually performed are different for different classes of material. This is due to limitations in the experimental apparatus rather than anything else. On the stiff materials such as bone it is usual to investigate the effect of speed of stretch applied uniformly over the entire stress–strain curve (or at least over a major fraction of it). On the more compliant materials the dynamic measurements are usually performed by applying length changes or loads which investigate the performance of only a very small region of the stress–strain curve.

This subject really becomes more complex than we can discuss here. The main points to be learnt are, however, that the magnitude of the elastic modulus does depend upon the speed of measurement, and that materials exhibit hysteretic effects. This introduces us to viscoelasticity.

5.10   Viscoelasticity

Many materials have fairly straight length–tension curves before the elastic limit. In particular artificial springs are very linear for considerable extensions. It is very convenient to define what is known as a *Hookean spring* as a theoretical entity as one which obeys Hooke's Law

$$\text{tension } (T) \propto \text{length } (l) \tag{5.16a}$$

The constant of proportionality is the *stiffness* (k), which is therefore a constant for this entity.

$$T = kl \tag{5.16b}$$

The extent to which biological materials are such springs can be seen by looking at their stress–strain curves and seeing how straight they are. Over the extreme range, obviously rather crooked. Over short regions they are often fairly straight. A further property of the Hookean spring is that the same stiffness is obtained under any imposed conditions. Such an element cannot show hysteresis. The only materials which has so far been shown to have this attribute to any great extent are the animal rubbers (see page 63).

Thus in order to begin to explain hysteretic effects in terms of simple elements we need another kind of element. This is known as the *dashpot* and is a viscous element (a plunger in a tin of treacle conjures up the right kind of mental picture). It has the property that the tension exerted (T) is proportional to the rate at which the length is changing (the velocity). Thus:

$$T = \beta \, dl/dt \tag{5.17}$$

in which β is termed the viscosity of the dashpot.

A considerable amount of effort has been exerted in trying to build models of materials composed of combinations of springs and dashpots. The two simplest, known as Voigt and a Maxwell element, are illustrated in Fig. 5.6.

The use of such elements does not really help understand what is going on inside any particular material. No really good understanding of the phenomena responsible for the viscous component of hysteretic effects exists in the form of the precise theories to explain the elastic effects. However they are thought to arise from the sliding of the molecules of the material past one another and the rearrangement of relative positions resulting from applying strains to disordered matrices of molecules.

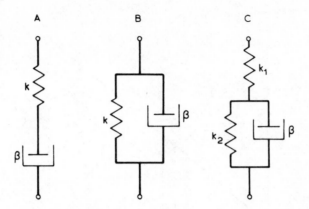

Fig. 5.6 A. Maxwell and B. Voigt elements of a single spring and dashpot. C. An extra spring in series with a Voigt element. Many materials have dynamic properties which are fairly well described by this network.

Nonetheless they are useful in gaining insight and a simple *description* of the behaviour. Many materials have similar viscoelastic properties to those obtained from the network of Fig. 5.6C.

5.11 Elastic elements as stores of potential energy

In order to extend an elastic element a force has to be applied. If we just consider Hookean springs for the moment, then this force will be proportional to the extension. We saw in Chapter 4 that work is done when the point of application of a force moves, and that the work is the product of the force and the displacement. We can readily evaluate the amount of work that must be done to extend a spring of stiffness k by an amount x. Fig. 4.7 shows the tension–length curve for the spring. The work done in extending the spring by a small amount dx when the extension is x (at which point the tension T is kx) is:

$$\text{work} = \text{tension} \times \text{displacement}$$

$$= T\,dx = kx\,dx \tag{5.18}$$

Notice that this amount of work ($T\,dx$) is the area of the slice under the tension–length curve between the extensions x and $x + dx$ as shown in the figure. The total amount of work required to extend the spring over the complete extensions X is found by integrating this expression over this extension.

$$\text{total work} = \int_0^X kx\,dx = \tfrac{1}{2}kX^2 = T^2/2k \tag{5.19}$$

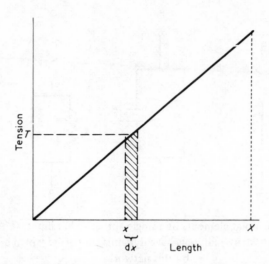

Fig. 5.7 Stress—strain curve of a Hookean spring. The work done in extending the spring by dx when its strain is x is the dashed area.

This is simply the area under the tension—length curve between these limits. In Chapter 4 we made the point that the potential energy stored in a system was the work performed in attaining that system.

Thus the potential energy of a stretched Hookean spring is $\frac{1}{2}kX^2$, and, more generally the potential energy of any elastic element is the area under the tension—length curve, whether this is linear or not.

5.12   Energy loss in hysteresis

We have pointed out that some materials, especially when stretched fairly rapidly, show hysteresis; their tension—length curves are different for increasing than for decreasing lengths. This means that the work done in stretching them is different from the amount of work obtained as they are released. Consider the case in Fig. 5.8. The work done in stretching is the area under the stretching curve (cross-hatched one way) and that obtained during the release the area under the release curve (cross-hatched the other way). The difference between these (equal to the area of the loop itself) is the energy lost in the specimen. Notice that the curve is traversed clockwise.

There is one very interesting biological material which exhibits what we can call negative hysteresis. This is insect flight muscle. In this muscle a loop like that shown in Fig. 5.8 is obtained if the muscle is stretched and released by small amounts at the right speed, with the important difference that the loop is traversed anti-clockwise (the arrows would have to be put on the diagram in the reverse direction). This means that less work is done in stretching the muscle than the muscle gives back when it is

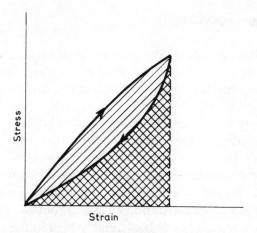

Fig. 5.8 Stress–strain curve of a material exhibiting hysteresis. The curve obtained for an increasing
applied length is above that for a decreasing length, as shown by the arrows.

released. Thus energy is being obtained from the muscle, and this is used to drive the
wings of certain orders of insects. How it does this is outside the scope of this book.

5.13 Isotropic and anisotropic materials

So far we have said nothing about the molecular properties of our materials. Without
saying anything about it we have been talking as though the elastic properties of our
specimens have been the same in all directions. This is often the case with non-
biological materials; less so with biological ones. For example a piece of steel has the
same properties independent of which direction it is being pulled in. This is not the
case with a piece of bone. Bone is much stiffer and has a greater tensile strength in
its longitudinal direction than transversely. We say that the steel is *isotropic* meaning
that its properties are the same in all directions, and that bone is *anisotropic* since
its properties are different in different directions. The degree of anisotropy in a
specimen must reflect the way in which it is made. Steel is made up of a network of
atoms which is symmetrical in all directions. Bone is much more complex, but is
basically a matrix of collagen in which crystals of apatite (calcium phosphate) are
embedded. These crystals are generally fairly long and very thin. The collagen is
made up of long fibres. Thus structurally bone has two materials each of which are
directionally differentiated. It is not surprising that the mechanical properties reflect
this structural directionality.

Thus when talking about the elastic properties of any material, care must be taken
first of all to establish whether or not it is isotropic and secondly, if it is not, to specify
the direction in which the measurements are made.

5.14   Short and long-range elasticity

What molecular mechanisms enable the great range of elastic moduli to be obtained?
Basically there are two, which are given a variety of names, since they exhibit a
number of distinct and different properties. We shall use the terms short-range and
long-range elasticity.

Short-range elasticity is the easier to understand. It is obtained in materials in
which there is a rigid lattice of molecules bound together by strong bonds. Examples
are metals and crystals. Given a material of this sort, the effect of an applied strain is
to try to distort these bonds, thereby setting up large forces within the material. The
experimental evidence for the way in which the potential energy between two
adjacent atoms changes as their separation changes is outside the scope of this book.
In general the stiffness of such bonds is extremely high in the region of equilibrium,
but falls off with distance. Thus very high forces are needed to cause the initial
separation, but once this is achieved the force is quickly reduced and the interaction
is 'broken'. Thus materials of this sort have high stiffness, and their breaking strain is
very low. The energy required to stretch the materials is stored in the material in the
form of potential energy in the bonds. It therefore is contributing to the internal
energy of the material.

One particular influence upon the stiffness of interest is the effect that temperature
changes will have. What will happen if we increase the temperature? In the case of
short-range elastic materials the thermal energy of the individual atoms making up
the material will be increased and this will enable the individual bond lengths to
become slightly longer. Although the position of zero force may not have been
effected by the temperature, the mean distance away from this position that the
atoms will travel is determined by their thermal energy; the greater this is, the
greater the separation can increase. Therefore, the material as a whole will expand,
and will, at any given length, exert a smaller tension.

Long-range elasticity is the characteristic of materials which as a class are known as
rubbers. (The term of course arises from rubber which is the best known material of
this class, although there are many biological examples of which resilin is the most
widely quoted). The characteristic structure of all this class is that they are made up
of what are called *long-chain molecules*. Proteins and nucleic acids are examples of
long-chain molecules. In general they are molecules which are composed of a long
sequence of atoms. These are often carbon-carbon sequences of one kind or another.
Thus the simplest example would be a high molecular weight paraffin without side
branches as shown in Fig. 5.9. Complexity can be achieved by having groups other
than hydrogen attached to the carbon atoms of the main chain, by having substantial
side chains and by inserting atoms other than carbon into the main chain, as is done,
for example, in proteins which are essentially long sequences of amino acids.

Although all rubber-like materials are composed of long-chain molecules, it does
not follow that all substances composed of long-chain molecules exhibit long-range

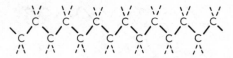

Fig. 5.9 Long carbon chain in extended configuration. For saturated bonds a carbon atom can
rotate fairly freely relative to its adjacent carbon atoms.

elasticity. There are two further essential features that have to be attained. The *first*
is that the molecules must not have a rigidly determined shape — the backbone must
be free to take up a wide variety of different shapes due to rotation about the bonds
of the backbone. Most proteins and nucleic acids do, of course, have very precise
structures. This is because of the precise interactions between the side groups and
between different parts of the backbone with each other (often in the form of
hydrogen bonds), and also because of the interaction of the side groups with the
aqueous environment. The rubber-like proteins do not have this precise orientation.
They are characterized chemically by having inert side-groups and by the presence of
proline which tends to prevent hydrogen-bond formation between regions of the
backbone. The *second* requirement of the rubber-like material is that the long-chain
molecules shall be *cross-linked* to one another to form a continuous matrix. A set
of independent molecules, such as is found in the paraffins, would provide no
mechanical continuity. Further the cross-links must be widely spaced apart on the
molecules — too much linkage will prevent the sections between the links being
able to take up a random structure. Thus all the rubbers have long-chain molecules,
cross-linked to form a continuous, but loose matrix with no other interaction
between the chains, and the sections between the cross-links being free to take up
random configurations. To a fair approximation, one atom of the backbone (or
group of atoms in proteins) can rotate freely relative to its neighbours. Thus a *very*
large number of different conformations of each molecule is possible. Let us, for
the moment, imagine our material to be made of just one long chain molecule. The
length between the ends of the molecule will depend upon the degree of folding of
the chain. In the one extreme the atoms of the backbone could be laid out in a long
line, in which case the overall length would be considerable. At the other extreme
the molecule could be tightly folded into a ball, and the overall length would be
very short. We can see why such extreme length changes are possible. (Fig. 5.10).
What determines precisely what the length will be at any moment of time? Surely
in fact the length will be constantly changing because of the thermal energy of
each of the constituent atoms. Suppose we think about each of the individual bonds
which determine the orientation of one element of the backbone relative to its
immediate neighbours. This can have one of a number of different orientations.
However, because the molecule really is rather long we find that at any instant, *on
the average*, there will be a certain number of bonds in one orientation and a certain
number in another and so on. In other words, on the average, the overall length of

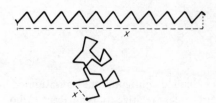

Fig. 5.10 Separation (X) between the ends of an extended (upper) and a randomly coiled (lower) chain.

the molecule will be fairly well determined simply because there will be so many different ways in which the molecule can fold itself to obtain this length. The extreme length, with all bonds orientated in the same direction, can only be achieved in one way. Thus the chances that all the bonds are simultaneously in this direction is very small, and the chances are that this shape will not be encountered. The number of different ways in which the average length can be obtained is very large, and thus this length will be that most frequently experienced. The reason why we are speaking in terms of the length continuously changing is because of the thermal energy of the constituent atom. Thus, suppose that we pull on the ends of the molecule, so that we distort it from its most probable length, there will be a force experienced and the source of this force will be the thermal energy of the atoms trying to make the molecule take up a more probable or likely set of bond angles. Because the molecule is so twisted, the effect of the pull on the lengths of the individual bonds will be very small indeed. In other words the contribution to the tension of the change in internal energy due to bond distortion is minimal. The source of the tension is the tendency to take up a more probable length due to the constant thermal energy in the material.

The theoretical relationship for the tension as a function of length in such a molecule is given by the very simple relationship:

$$\text{Tension } T = \alpha kT x \qquad (5.20)$$

in which the extension is x, kT is thermal energy (see Chapter 11) and α is a constant obtained from the expression for the probability of such a molecule having any particular length.

The molecule wants to take up its most disordered state. Pulling tends to provide order into the system. This is precisely what we mean by the entropy of the system (see page 100). The unstressed molecule has maximum entropy and the effect of applied strain is to reduce the entropy of the system. We say that the source of long-range elasticity is *entropic* for this reason.

What happens if we heat rubber? Suppose that the rubber is initially being held at some extended length. The effect of heating will be to increase the thermal energy, and thus to give a greater tendency to become disordered. The rubber will thus exert a greater tension or in other words will try to shorten. Thus if the rubber is extended by a constant load it will get shorter. Similarly if it is being compressed it will tend to extend to its original length. This is the opposite effect to that obtained with short-range elastic materials.

Although the above argument was given for a single molecule, the above reasoning still holds in principle for the cross-linked matrix that comprises the complete rubber. Instead of talking about the distance between the ends of the molecule, we are now considering the distance between adjacent cross-links. Obviously there will be a restriction on the shape changes imposed by the interaction between the different molecules, and this leads to modifications in the theoretical treatment. In natural rubber there are rather few cross-links, and artificial ones have to be introduced before tyres and other materials are made. The process of introducing cross-links is known as vulcanization, and is achieved with sulphur bonds. The zoological rubbers which are proteins have evolved their own special forms of cross-linkage. These rubber-like materials fit the theoretical formulae for rubbers better than natural rubber, and the reason is that nature introduced cross-links at regular intervals, whereas the vulcanization process introduces them at random.

The theoretical tension–length expression for the network is very similar, though slightly more complex, than for the single molecule. It is

$$T = NkT(\lambda - 1/\lambda^2) \tag{5.21}$$

where N is the number of cross-links per unit volume and λ is known as the extension ratio and is defined as

$$\lambda = \text{strained length/unstrained length.} \tag{5.22}$$

It is thus slightly different to strain and is in fact

$$\text{Extension ratio} = \text{strain} + 1 \tag{5.23}$$

5.15 Animal 'rubbers'

At least two rubber-like proteins are found in the animal kingdom, and it is likely that more will be discovered. These are resilin from insects and abductin which is found in bivalve molluscs, pushing the two shells apart. Resilin has been the better studied. It exhibits good elastic properties and shows minimal hysteresis. It has been discovered in the wing-hinges of locust and dragonflies and in the jumping legs of fleas.

Why do animals develop rubber-like materials? One reason is to act as an efficient store of mechanical energy. The resilin in the insect wing hinge is an excellent example. The flight muscles are used to make the wing move up and down. Since the wing must move very rapidly the force of the muscle is converted into kinetic energy of the moving wing. At the end of its stroke the wing must stop and reverse direction. Thus, all the kinetic energy has to be lost. In the absence of an elastic material the only way of stopping the wing would be to have a stopping muscle. The energy would be lost as heat. By using an elastic material however the kinetic energy is converted to potential energy in this material. The wing is slowed down against a spring which then helps to start it moving in the opposite direction. This is a great saving in the energy requirement of the flying insect.

5.16   Intermediate elastic moduli

So far we have only discussed two types of material. These are the short-range and the long-range elastic materials with Young's moduli of about $10^{11}\,\text{N/m}^2$ and $10^6\,\text{N/m}^2$ respectively. (See Table 5.1)

This is a fantastic difference in elastic modulus – a factor of 100 000. Animals require materials with intermediate values for a whole variety of purposes. How are these obtained? In fact it is not really the moduli of elasticity that concerns the animal but the stiffness and extensibility of a particular structural part. In principle it is possible to obtain structures of low stiffness from materials of high elastic modulus by using very small cross-sectional areas. This does not solve the problem however, because such materials can still only be stretched by very small amounts. Greater extensibility is needed. In fact animals do make use of very thin sheets of fairly stiff material for situations in which the material needs to be bent rather than stretched, as in the cuticle between segments and joints in arthropods.

Animals have developed a whole range of materials with intermediate elastic moduli. These are the fibrous proteins. Their man-made counterparts, from the point of view of stiffness are the polymer fibres that are so commercially successful (the nylons, rayon, polyethylenes, etc.).

How are elastic moduli in the range $10^7 - 10^{10}\,\text{N/m}^2$ achieved? One way is by being made up of long-chain molecules, as in the rubbers, but having a much greater extent of cross linking. The resulting matrix is more rigid than in the rubbers, because less rearrangement between cross-links is possible, but much less stiff than the crystals and the metals, because movements of chains relative to one another are still possible. Another way is to have what in effect is a composite material in which there are regions of very high elastic modulus, joined by regions of very low elastic modulus. The resulting modulus will be intermediate both in modulus of elasticity and in extensibility (possible strain before the elastic limit is reached). Such materials can be made from what is basically the same building block of long-chain molecules by having very extensive cross-linking in some regions (regions of crystallinity) and very loose cross-linking in other regions (regions of long-range elasticity). Alternatively the two regions can be different materials.

Man-made fibres tend to have a fairly uniform cross-linking, and make this more extensive the greater the elastic modulus that is required. Biological materials tend (as far as we know at present) to use the second method, and to have stiff and compliant regions. Thus the fibrous proteins tend to be made of aggregations of long thin molecules, whose individual structure is very precise and which has probably a fairly high elastic modulus. These molecules are aligned in the fibre. The forces holding the molecules together are not well understood, but are possibly the regions of lower elastic modulus.

5.17 Body walls and the hydrostatic skeleton

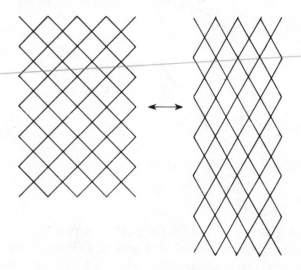

Fig. 5.11 A diagrammatic representation of the way in which a criss-cross pattern of fibres can lengthen (and become necessarily thinner) without length change of the fibres by a change of the angle of the pattern.

The extreme extensibility of certain animals was mentioned at the beginning of this chapter. How does the body wall cope with these changes? Not by having a highly extensible body wall in the sense of low modulus of elasticity. The animal requires a body wall which is sufficiently stiff that it will allow no significant volume change in the animal, but which can change its shape. It does this by having a criss-cross pattern of collagen fibres with an elastic modulus of about 10^8 N/m^2. The angle of the criss-cross can change, thereby allowing lengthening of the animal at a constant volume of its contents. This is shown schematically in Fig. 5.11. Any elastic restoring forces provided by the body wall in this case depend upon the way in which the collagen fibres are held together, not upon the elasticity of the collagen fibres. This is at one order of structure higher than the elasticity of the materials discussed in the rest of the chapter.

5.18   Questions

1. Evaluate the Young's modulus of the materials whose stress–strain curves are depicted in Fig. 5.5.

2. An insect muscle 10 mm long and 50 mm² in diameter, is found to have the stress–strain curves shown in the diagram when measured at the frequencies indicated.

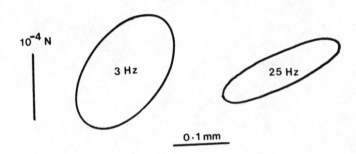

What is (a) the work per cycle and (b) the power output of the muscle at the different frequencies? (The curves are ellipses — the area of an ellipse is πab, where a and b are the long and short axes respectively.)

3. If the Young's modulus of bone is 10^{10} N/m² , what extension will be obtained when a 1 kg mass is hung from the end of a vertical peice of bone 10 cm long and with a square cross section of side 5 mm. If the bone is held horizontally from one end, and the 1 kg mass hung from the other end, what will be the vertical deflection?

4. Below the 'knee' at about 3% strain (Fig. 5.5), hair behaves as a short range elasticity; above the 'knee' hair behaves as a long range elasticity. What does this suggest as the nature of the molecular events occuring at the 'knee'?

5. If half the free energy of one molecule of ATP (40 kJ/mole) were used to extend an elastic molecule by 10 nm, what would be the stiffness of the elastic molecule?

6. What is the combined stiffness of two rubber bands of individual stiffness k_1 and k_2 (a) in series, (b) in parallel?

7. Two imaginary animals have skeletons made from the same material. One has a tubular endoskeleton and the other a tubular exoskeleton. If the cross sectional area of the skeleton in the legs of the animal are both equal to 3 mm² and the external radii of the skeletons are 1 mm and 5 mm, what is the ratio of (a) the flexural rigidity and (b) the strength to bending of the two skeletons?

6 The Motion of Fluids

This chapter deals with the interactions which take place when a body moves through fluid, such as a fish through water or a fly through air, or when fluid flows past a solid object, such as blood through blood vessels and air flowing through trachea. It is unimportant to know whether the fluid or the solid object is in motion — we are concerned only with the relative movement of the one with respect to the other.

We must first of all determine what types of force are involved, and shall deal only with the case when we have reached equilibrium, that is when the relative velocity between our fluid and solid is constant and there is no acceleration. When this is the case, then by Newton's third law of motion, the net force acting either upon the fluid or solid is zero. In the case of the fish swimming through water the force exerted by the fish in swimming must be balanced by the dragging forces tending to prevent movement. It is just these dragging forces that we are concerned with.

We must consider what we mean by a fluid, and determine what properties it possesses. For our present purposes, the molecular nature of matter treats the fluid in too microscopical a way. We do not want to consider the individual motions of molecules, but rather the overall macroscopic movement of large numbers of molecules. For this reason we can think of our fluid as being a continuum, and we will find it useful to consider a small volume of this continuum from time to time. We will call this small volume a fluid particle, and this particle will obey Newton's laws of motion. In particular it will possess a certain mass, and when acted upon by a force will be accelerated. It will also possess that property of fluids known as viscosity.

6.1 Viscosity

In the physics of solid elasticity (see Chapter 5) we defined the modulus of elasticity as the ratio:

$$\text{Elastic modulus} = \frac{\text{stress}}{\text{strain}}$$

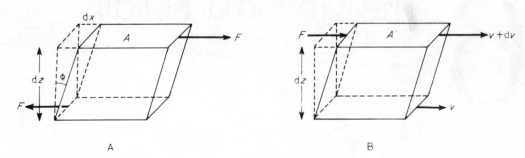

Fig. 6.1

In the case of a shearing force the modulus that is obtained is known as the rigidity modulus (n) which is defined using the symbols of Fig. 6.1A as

$$n = \frac{F/A}{\phi} = \frac{F/A}{dx/dz}$$

In the case of fluids it is not possible to define elasticity in this way because a constant shearing force does not produce a constant deformation. In this case, (Fig. 6.1B) a constant shearing force will produce a constant velocity gradient, perpendicular to the shearing force and we define the viscosity, or drag, between each layer of fluid as:

$$\text{Viscosity } (\eta) = \frac{\text{shearing stress}}{\text{velocity gradient}} = \frac{F/A}{dv/dz} \tag{6.1}$$

Thus in Fig. 6.1B, if a constant shearing force F is applied across a small volume of liquid of thickness dz, then the top surface will move relative to the bottom surface with a constant velocity dv. The magnitude of dv will be related to the magnitude of the applied force by means of equation 6.1.

The dimensions of viscosity can be derived from equation 6.1. Equating the dimensions on each side of the equation we obtain:

$$\text{Dimensions of viscosity } = \frac{ML}{T^2} \times \frac{1}{L^2} \times \frac{LT}{L} = \frac{M}{LT}$$

and in SI units are therefore kg/m s or Ns/m^2.

The viscosities of a number of fluids are given in Table 6.1.

The above is a phenomenological description of viscosity. At the molecular level the explanation of viscosity is in terms of the random movement of molecules in the direction perpendicular to the mass movement of the fluid. Consider a plane in the fluid perpendicular to the direction of the velocity gradient (e.g. the plane A in Fig. 6.1B). The molecules immediately above the plane have a slightly different speed in the direction of mass travel than those below the plane. Due to random movement

TABLE 6.1

Fluid	Viscosity η Ns/m^2 20°	Density ρ kg/m^3	η/ρ m^2/s
Water	0·0010	1·00 × 10^3	1 × 10^{-6}
Alcohol	0·0012	0·79 × 10^3	1·5 × 10^{-6}
Glycerin	1·49	1·26 × 10^3	1·18 × 10^{-3}
Olive oil	0·084	0·92 × 10^3	9·1 × 10^{-5}
Air	1·82 × 10^{-5}	1·29	1·4 × 10^{-5}

of the molecules however there will be a transfer of momentum across the plane, which will tend to equate the molecular movement above and below the plane, and will thus produce the shearing stress.

If the fluid is a gas, at a pressure not too far removed from atmospheric pressure, then the viscosity is independent of the pressure. The reason for this is that the mean free path (the average distance between collisions) of the molecules increases as the pressure is reduced. This means that, on the average, the molecules crossing the plane of the above paragraph will have come from a greater distance away from the plane before crossing it, the lower the pressure. Their momentum will therefore be that relevant to the fluid at a greater distance away, and will therefore be more different from that of the fluid in the vicinity of the plane being considered the lower the pressure. The transfer of momentum across the plane is thus independent of pressure — the lower the pressure the less the frequency with which molecules cross any particular plane, but the greater their difference in momentum. This only applies provided the mean free path is small compared with the overall dimensions of the fluid. In air at normal temperature and pressure, the mean free path is about 0·1 μm.

6.2 Reynold's number

The two aspects of a fluid particle that we need to take account of when considering its motion are its mass and its viscosity. An external force acting on the particle will have two effects. In the first place, the particle will be accelerated according to Newton's second law. The force resisting acceleration due to mass of the particle is known as the *inertial force*, and is a vector in the opposite direction to the acceleration, and of magnitude equal to the mass multiplied by the acceleration. In the second place, the external force will cause the particle to move relative to its neighbours, setting up a *viscous* shearing *force* acting against the applied force. The external force will be opposed by two forces therefore, the inertial and the viscous. It is important to know which of these two forces is the dominant one, because different phenomena occur in the two regions where either the inertial or the viscous forces are the more important. For this reason, when discussing a situation, the ratio of these two forces is often referred to, and is known as the Reynold's number (R_n)

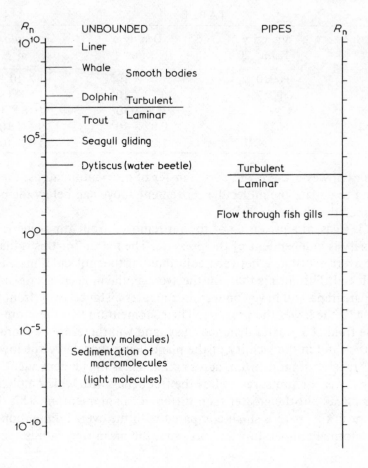

Fig. 6.2

for that situation. The Reynold's number is defined as:

$$R_n = \frac{\text{density} \times \text{velocity} \times \text{length}}{\text{viscosity}} = \frac{\rho v l}{\eta} \qquad (6.2)$$

and is equivalent to $\dfrac{\text{inertial force}}{\text{viscous force}}$

In the above equation, the density and the viscosity are those of the fluid, the velocity is the relative velocity between the fluid and the solid, and the length is a measure of the size of the solid. Thus in the case of fluid flowing through a pipe, it might refer to the radius of the pipe. In the case of a fish swimming through water, the total length of the fish would be suitable. It might seem surprising that there is no rule for what length to take. One of the main uses of Reynold's number is that it tells us when experiments are equivalent. Thus when testing aeroplanes in wind tunnels or boats in water troughs, the model situation will be exactly that of the final construction when the value of R_n for the two cases is the same. In this case the length chosen does not matter provided that the same dimension is chosen in the two cases. For our purposes we do not need to know the Reynold's number exactly, and any convenient length may be taken. It may also seem a little odd that when talking about forces acting on a fluid particle, a length measurement apparently unrelated to the particle is introduced. The reason for this is that we are discussing the fluid particle in reference to some situation, and it is necessary to scale the size of our fluid particle up and down in the same ratio as the dimensions of the objects in this situation, by they walls of a pipe or the body of a bird. It will be seen that too much importance must not be attached to the exact value found for R_n, we are simply getting an idea as to whether we should be thinking in terms of mainly viscous forces (when R_n is small) or mainly non-viscous, that is inertial forces (when R_n is large). The values of R_n for the movements of various bodies is shown in Fig. 6.2.

6.3   Fluid bounded and unbounded

There are two types of problem that we have to consider. The first is that presented when the fluid is bounded, as is the case with fluid flowing through pipes; the second is that which occurs when the fluid is unbounded, and we are discussing the flow of an infinite fluid around a solid object. For both cases we must consider the effect of low and high Reynold's number, that is, the effect when the viscosity is important or unimportant.

6.4   Flow of fluid through pipes — low R_n

At low values of R_n in the flow of fluid we have what is known as *streamline* or *laminar* flow, as opposed to *turbulent* flow. These two states can be observed in water flowing from a tap. When the water is coming out slowly the flow is smooth, and the path of any small volume of water follows that of the whole flow. If the tap is turned on further however so that the water is pouring faster, then the appearance of the flow is ragged and a great deal of buffeting about within the flow is evident. The path followed by a small volume of water is no longer smooth, but now transfers itself from one part of the flow to another, and we have turbulence. Whether we have laminar or turbulent flow in any particular situation depends upon the velocity of the fluid, and the value of the velocity when turbulence occurs is known as the *critical velocity* (V_c). For the flow of fluids down pipes the critical R_n is about 1000, and the critical velocity is then given by:

$$V_c = \frac{1000 \times \eta}{\rho \times a} \qquad (6.3)$$

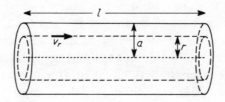

Fig. 6.3

where a is the radius of the tube.

Provided we have streamline flow, then we can determine both the velocity and the volume of fluid flowing down a pipe under a certain pressure. Consider fluid flowing down a length l of pipe of radius a, when the pressure difference between the ends of the pipe is P. Once a steady state has been reached no liquid will be accelerating. Further it is a property of the flow of fluids past solid surfaces that at the boundary between the two there is no motion of the fluid relative to the solid. Thus the rate of flow down a pipe is independent of the nature of the pipe, and depends only upon the viscosity of the fluid, the dimensions of the pipe and the pressure difference. Suppose the velocity of fluid at a distance r from the axis of the pipe is v_r, and consider the forces acting upon a cylinder of fluid of radius r.

Force acting on fluid = Pressure difference × area acted upon by pressure

$$= \pi r^2 P$$

Viscous drag = viscosity × velocity gradient × area under shearing force

$$= -2 \pi r l \eta \frac{dv}{dr}$$

At equilibrium the pressure force will equal the viscous drag and therefore:

$$\pi r^2 P = -2 \pi r l \eta \frac{dv}{dr}$$

Integrating

$$v = \int_0^{v_r} dv = \int_a^r \frac{-Pr}{2l\eta} \, dr = \frac{-P}{2l\eta} \left[\frac{r^2}{2} \right]_a^r = \frac{P}{4l\eta} (a^2 - r^2) \qquad (6.4)$$

This gives the velocity of flow of a cylinder of radius r. We can now determine the volume flow per unit time. In this case the volume flowing in a thin shell is determined and this result integrated over the total volume of the pipe.

Consider a cylindrical shell of radius r and thickness dr.

If the velocity of the shell is v and the velocity gradient across the shell is dv, then the volume of fluid per second passing through the shell is given by:

$$\text{Volume/second} = 2 \pi r \, dr \, v$$

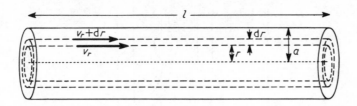

Fig. 6.4

The total volume of fluid per second passing through the tube is therefore

$$\frac{\text{volume}}{\text{second}} = \int_0^a 2\pi r\, dr\, v$$

The velocity v of the cylinder was evaluated above (equation 6.4). Substituting for v:

$$\frac{\text{volume}}{\text{second}} = \int_0^a \frac{2\pi P}{4l\eta}(a^2 - r^2)r\, dr = \frac{\pi P}{2l\eta}\left[\frac{r^2 a^2}{2} - \frac{r^4}{4}\right]_0^a$$

$$= \frac{\pi P}{2l\eta}\left(\frac{a^4}{2} - \frac{a^4}{4}\right) = \frac{\pi P a^4}{8l\eta} \tag{6.5}$$

Thus, as would be expected, the rate of volume flow of fluid is proportional to the pressure difference across the ends of the tube, and is inversely proportional to the length of the tube and the viscosity of the fluid. The dependence upon the radius of the tube however is very striking, varying as the fourth power of the radius. Thus to halve the flow, the radius needs to be reduced by only 16%.

This last fact is of particular importance in the vascular system. In the first place the large arteries and veins offer comparatively little resistance to the passage of blood; whereas the capillaries account for by far the greatest amount. Further the rate of blood flow will be strongly dependent upon the size of the blood vessels. The distribution of blood to various parts of the body is controlled by smooth muscle in the walls of the arterioles, and a reduction in diameter to a half the former value will reduce the blood flow by sixteen times.

Equation 6.5 is known as Poiseuille's formula, after the French physiologist Poiseuille (1799–1869) who was the first person to investigate the flow of fluid through tubes, and who gave his name to the unit of viscosity. It demonstrates one way of measuring the viscosity of fluids, and one that is very widely used by biologists. Measurement of the viscosity of solutions of known concentrations is one of the ways used to determine the shapes and sizes of macromolecules. Viscosities are generally measured with an Ostwald viscometer (Fig. 6.5A). A known volume of solution is used, and the time taken for this volume to flow through the tube BC is measured. In principle the absolute viscosity of the solution could then be determined, but in general the viscometer is calibrated by making a second measurement on pure

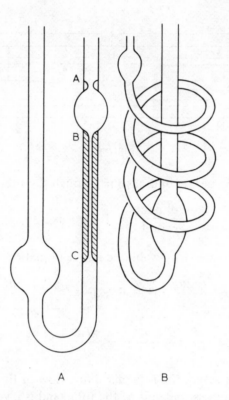

Fig. 6.5 Two types of viscometer used in biological work.

solvent and making comparative measurements. For accurate work the time taken
for the fluid to flow down BC needs to be as large as is practical. This means that
one needs a long tube with a small radius. However a small radius means that the
effect of impurities such as dust particles is maximized, and also that the shearing
forces acting on the fluid are large, which leads to spurious results when asymmetrical
molecules make up the solution. A more accurate, yet still manageable viscometer
of the form shown in Fig. 6.5B is therefore often used, which has a longer tube of
larger radius than an Ostwald viscometer.

The flow of fluids through pipes at large Reynold's number where the flow is
turbulent is a complicated subject and unimportant for our purposes, and so will not
be discussed.

6.5   Flow of an unbounded fluid past solid objects — low R_n

Reynold's number was defined as $\dfrac{\text{density} \times \text{speed} \times \text{length}}{\text{viscosity}}$

The region under discussion in this section therefore will be that where the density, velocity and size is small, and the viscosity large. This will include the motion of macromolecules and microorganisms, flight in the upper atmosphere where the density is low, and.the movements of viscous fluids such as molten lava and glacier ice, together with a host of more normal situations, where the parameters of Reynold's number make for a low value of this number.

The viscous drag upon a body in this region can be found by the method of dimensions. The resisting force of a body in a fluid will depend upon:

1. The viscosity of the fluid.
2. The velocity of the fluid relative to the body.
3. The size of the body.

but will not, in the region of low Reynold's number, depend upon the density. Thus we can write:

$$F = k v^x \eta^y l^z$$

where k is a constant and x, y and z are coefficients whose value we wish to determine. Now the dimensions on either side of an equation must be identical, so:

$$M L T^{-2} = k (L T^{-1})^x (M L^{-1} T^{-1})^y (L)^z$$

Equating coefficients of:

$$\text{Mass} \qquad 1 = y$$
$$\text{Length} \qquad 1 = x - y + z$$
$$\text{Time} \qquad -2 = -x - y$$

from which we obtain

$$x = y = z = 1$$

Thus:

$$F = k v \eta l \qquad\qquad (6.6)$$

In the case of a sphere of radius a, the constant k is 6π and we have:

$$F = 6\pi v \eta a \qquad\qquad (6.7)$$

For other shapes the value of k is different. Volume for volume the frictional resistance to motion is least for a sphere and greater for any other shape, depending upon the surface area of the body. This, it should be noted is true only in regions of low R_n below unity. Equation 6.7 is known as *Stoke's law*, and is used for instance to determine the resistance to movement of macromolecules moving under large accelerations in solution in a centrifuge. Thus one method used to determine the shape and size of a particle such as a macromolecule is to determine the velocity of the particle in a solution under a large force. An experiment of this kind is known as a sedimentation velocity experiment. When a force is applied to a body the body will accelerate to that velocity where the retarding forces at that velocity just equal the accelerating force. This velocity is known as the *terminal* velocity (v_t) and is reached very quickly in most situations. In the case of a macromolecule in solution, the applied force is centrifugal. The solution containing the macromolecule is placed in a centrifuge (one capable of very high revolutions per minute is needed for macromolecules, since the applied forces must be greater than those giving rise to thermal motion) and the velocity of the molecules is measured by measuring the rate of movement of the boundary between pure solvent and solution using schlieren optics (a technique, not discussed in this book, for detecting changes in refractive index of materials).

When the terminal velocity of the molecules has been reached, then:

$$\text{Applied force} = \text{retarding force}$$

$$m \, \omega^2 \, R \, = \, k \, v_t \, \eta \, r \tag{6.8}$$

where m is the apparent mass of the molecule in the liquid, ω is the angular velocity, R the radius of the centrifuge arm, and the retarding force is given by equation 6.6.

The terminal velocity is thus dependent upon the size, shape and density of the particle. Although it is not possible to determine all three from this one experiment, this method in conjunction with others can provide the required information.

6.6 Flow of an unbounded fluid past solid objects – high R_n

We now come to a region which is rather more difficult to treat mathematically, because, except for a few simple cases, the flow of the fluid is not uniformly laminar. The region we are about to discuss is the one of fish swimming in water and birds flying in air, and includes all those situations where Reynold's number is greater than about 100. In most cases it will be much larger. We shall only consider passive bodies such as birds gliding, since the complications introduced in the treatment of flapping flight, and the active swimming motions of fish are too great.

The law of drag in this region, equivalent to Stoke's Law in regions where R_n is small, can also be derived by the method of dimensions. In this case, since we are in the region of large R_n, we assume that the viscosity is negligible, and that the drag is dependent upon the velocity of the fluid, the size of the solid and the relative velocity between the two.

Thus:

$$F \, = \, k \, v^x \, \rho^y \, l^z$$

Equating dimensions gives:

$$M \, L \, T^{-2} \, = \, k \, (L \, T^{-1})^x \, (M \, L^{-3})^y \, (L)^z$$

Equation the coefficients of mass length and time we find

$$x \, = \, z \, = \, 2 \quad \text{and} \quad y \, = \, 1$$

this is

$$F = k\,v^2\,\rho\,l^2 \qquad (6.9)$$

6.7 Bernoulli's equation

Consider what happens to a fluid as it passes through a constriction, such as that illustrated in Fig. 6.6A.

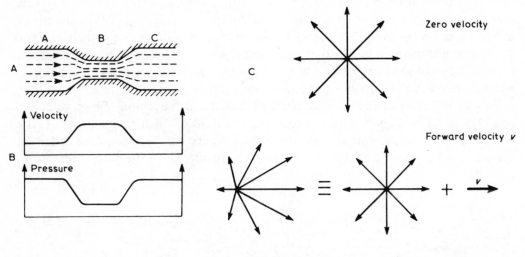

Fig. 6.6

Within the constriction the velocity of the fluid must be greater than outside, since the same mass of fluid must pass each point per second (the velocity of water passing through a ravine is much greater than the velocity in the wide open river on either side). This results in a reduced pressure of the fluid within the constriction. The reason for this can be explained in three ways: by considering what is responsible for the increase in velocity, by considering the molecular movements and by applying the conservation of energy.

Application of the principle of the conservation of energy gives rise to Bernoulli's equation, which is obtained by adding the kinetic and potential energies and setting the sum equal to a constant. For an incompressible fluid (liquid):

$$p\,V + \tfrac{1}{2}m\,v^2 \;=\; \text{Constant } (C)$$

where we are considering a fluid particle of mass m, volume V, pressure p and velocity v. Writing $m/V = \rho$ the density of the fluid, and dividing by V we obtain:

$$p + \tfrac{1}{2}\rho v^2 \;=\; C/V \;=\; \text{constant } H \qquad (6.10)$$

since for an incompressible fluid V is constant. In the case of compressible fluids we obtain a similar relationship.

$$\frac{\gamma}{\gamma-1}\, p + \tfrac{1}{2}\rho v^2 \;=\; H \tag{6.11}$$

where γ is the ratio of the specific heats at constant pressure and constant volume. For air γ is approximately 1·4.

Consider the molecular movements at the different velocities. At low velocities the direction of movement will be approximately random. The pressure, which is due to molecular bombardment, will thus be fairly high. However, if the molecules, on the average, are moving rapidly in one direction, their velocities in the other directions must be less, on the average, in order for the total energy to be the same in the two cases. This is illustrated in Fig. 6.6C. The static pressure is dependent upon the root mean square of the velocity (see Chapter 9) in the random directions, and is thus less when there is a mean flow in a particular, non-random, direction.

We can also see that the pressure must be reduced in the region of high velocity by considering the force necessary to accelerate the fluid to the higher velocity. This can be done only if the pressure is least where the velocity will be greatest, in order that the accelerating force (due to the difference in pressure) is in the correct direction.

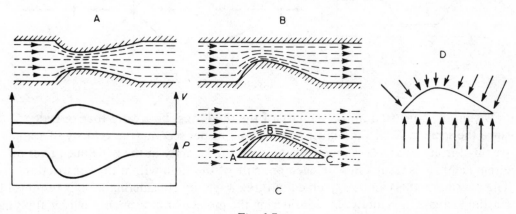

Fig. 6.7

Thus if we have a non viscous fluid flowing in a laminar fashion through parallel plates of the shape shown in Fig. 6.6A, the velocity of fluid in the neck region B will be greater than that in the wider regions A and C, since the volume flow through the two regions is the same. Thus the pressure of the fluid at B will be less than that at A and C. This is demonstrated graphically in Fig. 6.6B.

The constriction of figure 6.7A is similar to that of 6.6A, but with a slightly different shape. The velocity and pressure curves are as shown. Suppose that we have the constriction shown in Fig. 6.7B. The height of the constriction is at all points the same as that in Fig. 6.7A, and thus the velocity and pressure curves are the same.

Now consider the effect on the object of Fig. 6.7C (which has the same shape as the constriction, in Fig. 6.7B) isolated in a moving non-viscous fluid. The fluid immediately above the object will be compressed by the fluid further above, and the velocity profile will thus still be similar to that shown in Fig. 6.7A. The flat surface below the object will lead to little change in the velocity of fluid below, compared with the velocity of the bulk fluid. The net pressure acting at each point on the object will thus be as shown in Fig. 6.7D. There will be a net effect tending to make the object move upwards, due to the reduced pressure acting on the upper surface. There is thus a force acting on the object tending to produce lift. This is the principle of lift in the animal or aeroplane wing.

However, the problem has so far been too simplified, because even in regions of high Reynold's number, fluids cannot be treated as non-viscous in the region of solid objects, and this means that the infinitesimal layer of fluid in contact with the solid will be stationary, and there will be a thin layer next to this, known as the *boundary layer* across which the velocity is increasing from zero to the velocity of the streamlines of the fluid. This layer will be very thin, but the gradient of velocities across it will be very large, and hence the viscous forces will be large. The effect of these viscous forces will be to produce a dragging force on the object. This we will call *viscous drag*.

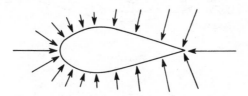

Fig. 6.8

There is however a second and more subtle form of drag on objects at high Reynold's number. This is known as *pressure drag*. We have seen how the pressure varies over the surface of an object due to the compression of the streamlines. This pressure will act along the normal to the body surfaces at each point, (Fig. 6.8) and the vector sum of the pressure summed over the surface of the body gives the net pressure force. If for some reason the pressure over the rear half of the body, which is that tending to cause forward motion, is not so great as that over the front half, then there will be a net pressure drag. In the case of a non-viscous fluid the pressure drag is zero. So far, in all the figures, an object moving in a fluid has been represented by a streamline shape. This is an object designed to minimize the pressure drag, and while considering pressure drag it will be more instructive to consider some other cross section.

In order to understand the onset of pressure drag, we must consider the movement

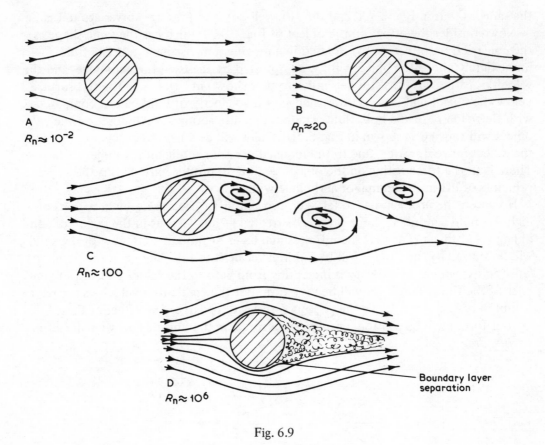

Fig. 6.9

of fluid particles in the boundary layer. These are subjected to three forces:

1. The accelerating force of the fluid outside the boundary layer.
2. The retarding force of fluid nearer the object.
3. A force due to the pressure gradient over the surface of the object.

Over the rear half of the body the pressure gradient tends to prevent motion of the fluid particle, and if the pressure gradient is large enough the boundary layer may be brought to a halt, and in extreme cases may even move in the reverse direction towards the front of the object. In either case the flow of fluid will be disrupted, and we obtain what is known as *boundary layer separation*. This is illustrated in Fig. 6.9. The effect of this is to reduce the pressure on the rear half of the object and give rise to a pressure drag (Fig. 6.10). The energy that is lost in this process is used in generating the eddying wake. With objects such as a wire the pressure drag is several times that of the viscous drag. As we have suggested, pressure drag is minimized by using an object with a streamline form. This has a reduced pressure gradient over its rear half, and hence boundary layer separation either does not occur, or when it does

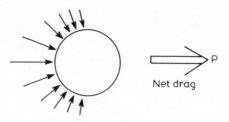

Fig. 6.10

occur, only does so at the rear end of the object.

Fig. 6.11 demonstrates the importance of the two types of drag (viscous and pressure drag) for two different cross sections, a circular and a streamline cross section. The sections have been chosen of such a size as to have equal total drags, and the relative sizes are also illustrated in the figure. The effect of streamlining is so great that the width of the streamline body at the point of maximum thickness is nine times that of the circular cross section. The situation at high R_n is very different from that at low R_n where the total surface area is the parameter that gives an indication as to the magnitude of the drag.

The streamline shape is found repeatedly in animals whose movements take place at high R_n.

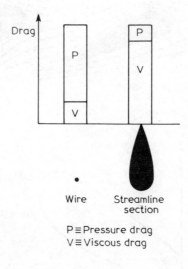

Fig. 6.11

6.8   Drag in laminar and turbulent flow

When discussing the flow of fluids down pipes we saw that above a certain value of

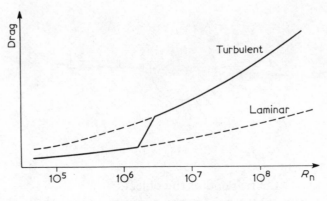

Fig. 6.12

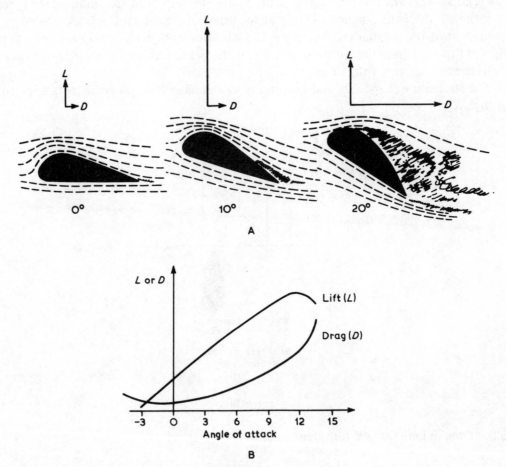

A

B

Fig. 6.13

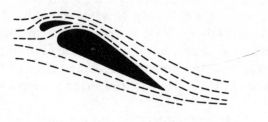

Fig. 6.14

R_n (about 1000), the flow became turbulent. The same effect occurs in the flow of
an unbounded fluid past a body, but in this case the value of R_n when this occurs
is about 3×10^6 for a smooth body. The relative viscous drag on a body as a function
of Reynold's number is shown in Fig. 6.12. Thus rigid bodies moving in a fluid where
R_n is greater than 3×10^6 will experience turbulent flow. The dolphin, however,
which swims with a velocity making R_n have a value of about 2×10^7 swims without
introducing turbulence. This is thought to be due to a rubbery outer layer of its skin,
whose effect is to dampen out any turbulence, thereby maintaining laminar flow.

6.9   The design of wings

The mechanism of lift of an aeroplane or animal wing was mentioned previously, and
illustrated in Fig. 6.7. A more suitable shape for the cross section of a wing is
illustrated in Fig. 6.13, and the net lift and drag components acting on the wing
when it is inclined at varying angles to the wind direction (*angles of attack*) are also
shown both diagrammatically and in graph form. The greatest lift occurs somewhere
in the region of 12°.

However at angles only a little larger than this the drag acting on the wing increases
rapidly due to the onset of boundary layer separation, with its ensuing eddying wake,
and production of *stall*. Thus the optimum angle of attack of a wing will be somewhere in the region of 6–10°. In order to achieve a greater angle of attack before
the onset of stall, some birds have evolved what is known as the slotted wing (Fig.
6.14). The effect of the leading section, whose angle of attack is several degrees less
than that of the main wing section, is to prevent boundary layer separation by causing
laminar flow over the top surface of the wing when the angle of attack is such that
stall would otherwise have occured.

Another problem with wings concerns the situation at the wing tip. The pressure
above the wing is less than that below it, and thus there will be a tendency for the air
to flow from below to above. The overall effect, due to the forward motion of the
wing relative to the air, is the setting-up of vortices at the wing tip, which increase

the drag on the wing; the drag is known as *induced drag*. The induced drag is most
pronounced at low speeds. It is obviously desirable to reduce this effect. For a wing
of a given area, the induced drag is least with wings that are long and narrow (such
wings are said to have a high *aspect ratio*, which is the ratio of the wing's length to
its 'chord', the distance from the front of the wing to the back). Thus birds which
spend a high percentage of their flights soaring (such as seagulls) have a high aspect
ratio.

However, animals which tend not to soar, but perform flapping flight rather than
gliding use less energy in beating their wings if these have a low aspect ratio. This is
because the inertia (see page 38) of the wing is less. Moreover a lower aspect ratio
leads to greater manouverability (which is why powered-flight planes have low aspect
ratios). In order to reduce the induced drag such birds splay out their primary feathers
at the wing tips. Instead of a broad wing tip the end of the wing is now broken up
into several smaller wing tips. The effect of this is to reduce the wing-tip vortices
and thus the induced drag. Since induced drag is lowest at slow speeds, the splaying
out of the wings becomes most pronounced at low flight speeds, such as when the bird
is coming in to land.

6.10   Pressure changes in the boundary layer

We have seen that near the surface of an object moving relative to the fluid surrounding
it, there is a boundary layer. At the surface the relative velocity between fluid and
solid is zero, and the relative velocity increases away from the surface. We have
further seen that the pressure at a point in a fluid is dependent upon its velocity.
However, there are *no* pressure changes across the boundary layer due to the differences
of velocity at different depths within it. Bernoulli's equation was derived from the
conservation of energy, and only applies to differences of pressure measured along
what is known as a *streamline*. A streamline, for laminar flow, is the path taken by
a fluid particle. In the boundary layer the motion of the fluid is parallel to the surface,
and there is no motion of the fluid particles perpendicular to the surface from regions
of one velocity to another. The conservation of energy still applies in the boundary
layer of course, but the viscous drag, responsible for the boundary layer, causes extra
terms to be necessary, not considered when deriving Bernoulli's equation. Energy is
lost in the boundary layer in the form of heat. Thus there will be no difference in
pressure across the boundary layer explainable in terms of the conservation of energy
due to differences in velocity induced by viscous drag.

However, viscous forces can still act in the boundary layer to produce effects
similar to those caused by pressure differences. Consider the situation illustrated in
Fig. 6.15 in which a U-tube has its ends at different depths in the boundary layer.
The fluid moving past the end of the U-tube will tend to drag out fluid from the U-
tube, and this tendency will be greater at higher velocities. Thus there will be a net

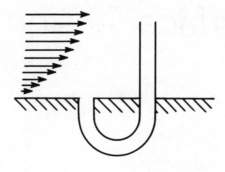

Fig. 6.15

movement of fluid in the U-tube due to viscous drag. This principle is used extensively by burrowing animals, both aquatic and terrestrial, for ventilating their burrows.

6.11   Questions

1. Your heart pumps blood at a rate of about 5 litres/minute at rest into the aorta which has a radius of about 1 cm. Assuming that blood has a viscosity of 4×10^{-3} Ns/m^2, and a density of 10^3 kg/m^3, what is a) the velocity of blood through the aorta? b) the critical velocity for laminar flow?
   What pumping rate of the heart would cause the critical velocity to be exceeded?
2. You have about 25×10^6 capillaries about 8 μm in diameter, and 1 mm long. What is the velocity of blood in the capillaries? What is the pressure difference expected across a capillary, using the data given in question 1, assuming that the viscosity of blood remains constant?
Note. In fact the corpuscular nature of blood results in an anomolous low viscosity of blood when measured in tubes whose diameter is comparable with that of the corpuscles.
3. What is Reynold's No. for a) a locust? b) a midge when gliding in normal flight? Assume sensible values for parameters you need. Can these animals fly using normal aerodynamic forces?
4. i) A macromolecule, with a molecular weight of 43 000 Daltons is spherical with a radius of 2·5 nm. Assuming that it has no water of hydration what frictional force would resist its motion if its velocity in water were 1 μm/sec?
ii) What centrifugal acceleration would be required to equal this frictional force? In a centrifuge, at a radius of 10 cm, what frequency of revolutions/sec would give rise to this acceleration?

7 Surface Tension

7.1 Surface energy and surface tension

The phenomenon of surface tension is one that occurs wherever there is a boundary between two substances. It is normal to consider surface tension as a property of liquids, and to talk about the surface tension of a liquid at an interface between that liquid and a solid, another liquid or a gas; however if we discuss the phenomenon from the point of view of surface energy, then we should also consider the surface energy of solids, and in particular the energetics of breaking solids.

Biologically the subject is of importance wherever a boundary occurs. Thus surface tension must be considered in discussing the operation of the lung, in which there is a boundary between the air and the fluid within the alveoli, and in the analogous organs in other animals, such as the insect trachea. The boundary between the water and the air at the surface of a lake is of relevance to a number of organisms which live immediately above or below the surface, such as pondskaters which skim over the surface or mosquito larvae which suspend themselves from the surface whilst they breath. A number of arthropods take a bubble of air with them when they submerge themselves in a pond, and use this as a temporary supply of oxygen; the continuing existence of this bubble depends upon the surface properties of the interface between the air and the water.

The surface properties of substances are due to the asymmetry of the forces between molecules at a surface. Within the body of a liquid or solid a molecule is surrounded by other similar molecules, and the average forces acting on it are equal in all directions. At the surface this is not true. If the surface under consideration is a boundary between a liquid or solid and a gas — say air — then a surface molecule experiences forces from one side only. The consequence of this is that it experiences a resultant force inwards (into the body of the substance). If the surface is a boundary between two liquids, then the surface molecule will be acted upon by dissimilar forces from either side of the boundary, and will experience a resultant force into one or other of the two liquids. For these reasons potential energy is associated with a surface, which we shall call the surface potential energy, or *surface energy*, and the magnitude of this is proportional to the area of the surface. For any particular substance the energy per unit area is a constant with units of $J\,m^{-2}$. The surface tension of water at a water/air interface is $7\cdot3 \times 10^{-2}\,N\,m^{-1}$. The surface area of a substance will thus tend to minimize itself in order to minimize the contribution to the potential energy from the surface. If the substance is a liquid then this can be achieved by the liquid taking on the shape with the minimum potential energy – however, since there are other contributions to the potential energy of a liquid, for example, the higher parts of the liquid have a greater positional potential energy than the lower parts, the shape taken by a liquid will not always be spherical (the surface with the minimum area for a given volume). Thus, raindrops, which are small, are spherical, since the contribution to the potential energy from the position of the water within the raindrop is small compared with the surface potential energy of the water composing the raindrop. However, a large drop of mercury on a table will have a flattened top and bottom, since in this case the contribution from the positional potential energy is of comparable magnitude to that from the surface potential energy. In space, where the positional potential energy due to gravity is negligible, even a large mercury drop is spherical.

In solids, of course, there is not the same freedom to change shape, because the cohesive forces between the molecules are so much stronger. Thus solids are not able to change their surface area so as to minimize the surface potential energy. This does not mean that there is not a surface energy in solids. One situation in which the concept of surface energy in solids is of considerable use is when considering their fracture. Work must be done to break a solid, e.g. a bone. The work done on the bone must obviously be reflected as a change in energy in the system after fracture – part of the change in energy in this case is the increase in the surface energy due to the creation of new surfaces. (The reason why the two pieces cannot then be re-united and have the same strength as previously is that the slightest change of shape in the surfaces after fracture is sufficient to create enough misalignment to prevent the molecules returning to the close proximity necessary for the inter-molecular forces to act.)

From now on we shall be considering the surface properties of liquids. We have so far considered the surface properties of the liquid in terms of a surface energy, of units $J\,m^{-2}$. However, a more satisfactory concept for most purposes is to think of the properties of the surface in terms of a *surface tension* of units $N\,m^{-1}$. Notice that the units $J\,m^{-2}$ are $(N\,m)\,m^{-2} = N\,m^{-1}$. The surface tension is to be considered as a force or tension acting parallel to the surface of the liquid, and of magnitude proportional to the length of the surface being considered. Thus for a surface of length l, and surface tension T, the total tension is Tl. The relationship between surface tension and surface energy is straightforward. If an area of surface is increased, for example as in Fig. 7.1 by moving a straight boundary of length l by an amount x, then the work done is equal to the force times the distance moved. The force is Tl, and the distance moved is x. Thus the work done is Tlx. This equals the change in energy due to the creation of an extra surface of area lx. If the surface energy of the

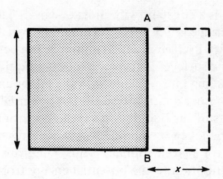

Fig. 7.1 Work done in increasing the surface area of a liquid. The side AB of the surface indicated experiences a tension Tl directed to the left. If the boundary is moved a distance x to the right, then this requires an amount of work equal to Tlx. (This equals the increase in surface energy due to an addition of surface area lx, with surface energy per unit area T.)

liquid is $T\,\mathrm{J\,m^{-2}}$, then the change in energy is indeed Tlx.

7.2   Angle of contact

At the edge of an interface between a solid and a liquid where both are in contact with a gas (as is obtained for example with water in a beaker at the side of the beaker, or at the edge of a drop of liquid on a solid surface) the net force acting on the liquid molecules is no longer directed perpendicular to the main surface of the liquid. The effect of this is to cause there to be an *angle of contact* between the liquid and the solid (see Fig. 7.2).

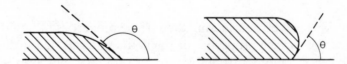

Fig. 7.2 Angle of contact. A drop of liquid on a solid surface has an edge making a specific angle θ with the solid as indicated. θ is known as the angle of contact. Two examples are shown, with angles of contact less than and greater than $90°$.

For this angle of contact to be realized the surfaces must be absolutely clean; any impurities can cause large changes in the angle of contact. Obvious examples of two very different angles of contact are those of a water drop or a mercury drop on a plate of glass. The angle of contact between water and glass is $0°$, that between mercury and glass is greater than $90°$.

For aqueous solutions in contact with other substances the direction of the resultant net force determines whether the other substance is described as being hydrophobic (water fearing) or hydrophylic (water loving). With hydrophobic substances the angle of contact will be greater than 90°, and with hydrophylic substances it will be less than 90°.

7.3   Surface energy of two liquids in contact

For two liquids in contact with one another the interfacial surface energy is the difference between the surface energies of the two liquids separately when in air. This relationship is only strictly true provided the surface energy of the individual liquids in air is measured under conditions in which the one liquid is saturated with the other.

One situation in which surface energies are of importance concerns the barrier between a liquid layer and an aqueous layer – e.g. between membranes and the solutions on either side of the membrane.

7.4   Curved surfaces

Curves surfaces are common. They are found in bubbles and drops, and also at liquid boundaries in thin tubes (due to the fact that there is an angle of contact between the liquid and the solid).

A curved surface can only be obtained when there is a difference in pressure between one side of the surface and the other. The action of the pressure in producing the curvature is balanced by the surface tension forces. Thus the pressure inside a bubble is greater than that outside.

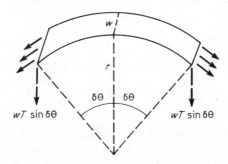

Fig. 7.3 Pressure inside a curved surface. The diagram shows a small width w of a cylindrical surface of radius r, subtending an angle $2\delta\theta$. The relevant surface tension forces, and their components inwards of magnitude $w\,T\sin\delta\theta$ are shown.

We can determine the relationship between the radius of curvature of the curved surface and the pressure necessary for its maintenance by reference to Fig. 7.3. The surface here is a width w of a cylindrical surface, of radius r, and we shall consider the forces acting on a small section of this surface subtending an angle $2\,\delta\theta$, as shown. The force acting outwards is due to the pressure difference between that inside and the outside. Pressure is force per unit area, and thus the force acting on the small segment is equal to the pressure times the area of the segment;

$$\text{Force outwards} = P \times 2\delta\theta\,rw = 2rPw\delta\theta \tag{7.1}$$

The inwards force acting on the segment arises from the component of the surface tension directed inwards. Now the surface tension acts parallel to the surface, and is of magnitude equal to the surface tension of the substance times the length of the boundary. Only the side of width w has a component acting inwards, and the magnitude of the inward component is

$$\text{Force inwards} = 2wT\sin\delta\theta = 2Tw\delta\theta \tag{7.2}$$

The factor 2 arises because there is a contribution from each end, and we are making use of the fact that provided θ is small, $\sin\theta = \theta$. At equilibrium the inward force must just balance the outward force, giving;

$$2rPw\,\delta\theta = 2Tw\delta\theta$$

Cylindrical surface: $$P = T/r \tag{7.3}$$

If the surface is spherical, rather than cylindrical, so that there is curvature in both directions, then a similar argument shows that

single spherical surface $$P = 2T/r \tag{7.4}$$

These two formulae refer to the situation where just one surface is involved (as in drops, or a bubble of air in water). Soap bubbles in air have two surfaces, and thus for a spherical bubble

spherical bubble (2 surfaces) $$P = 4T/r \tag{7.5}$$

Notice that there is a larger difference in pressure across the surface of a smaller bubble.

7.5   2 bubbles in contact — a problem for the lung

If two bubbles of different diameters are blown onto the opposite ends of a glass tube (as can be readily demonstrated with the apparatus shown in Fig. 7.4A) and then connected together, what will happen? Think of the answer before reading on.

Since the pressure inside the smaller bubble will be greater than that inside the larger bubble, the small bubble will send air to the larger bubble. Thus the small

(A)   (B)

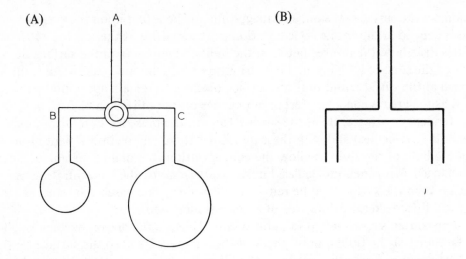

Fig. 7.4(A). Apparatus for demonstrating differences of pressure inside bubbles of different sizes. The tap at the junction of the tubes can be turned to connect any two of the tubes. It is first set to connect A and B, and the bubble on B is blown; next it is set to connect A and C, and the bubble on C is blown. Finally B and C are connected to see what happens. (B). Normal T-tube.

bubble will blow the large bubble up further, and in doing so will collapse. Conversely, if you try and blow two bubbles simultaneously using a T-tube as shown in Fig. 7.4B, you will be unable to do so. One side will initially form a bubble, and this bubble will be blown up with no air passing to the other side of the tube. Only the side which starts forming the bubble will be able to form one.

This phenomenon creates a crucial problem for the lung. In the lung, the regions in which the air and the blood are in extremely close proximity (the alveoli) are very small spherical regions. During breathing these are blown up by the incoming air, and then collapse (partially) as the air is expired. The alveoli differ in size. The walls are composed both of a thin membrane, and of a layer of fluid, and the elastic properties of the alveoli depend upon the mechanical properties of both the membrance and the fluid. The relevant properties of the fluid are the surface tension effects. Since the radius of the alveoli is so small, it is found that the surface tension of the alveolar fluid can account for up to 75% of the elasticity of the alveoli. The problem concerns the way in which the smaller alveoli can be filled, since from the above reasoning if air is blown into bubbles of different size, the air will pass entirely into the larger ones, and furthermore, the smaller ones will have a greater pressure, and so will collapse. Why don't the small alveoli collapse at the expense of the larger ones?

The answer to this problem concerns the effects upon the surface tension of substances in the aqueous fluid bathing the alveoli. The effect of impurities in a liquid is in general to lower the surface tension of the pure liquid. Some substances have very marked effects. In water the presence of detergents causes very large reductions in the surface tension. The magnitude of the effect upon the surface tension is dependent upon the concentration of the additive at the surface. Most impurities distribute themselves between the bulk of the fluid and the surface; if the area of the surface is changed, a redistribution of the impurity allows the concentration at the surface, and thus the surface tension, to remain approximately constant as the area is changed. Certain impurities on the other hand have a very marked tendency to aggregate at the surface (due to the presence of hydrophobic regions in the molecules) and with these molecules there will be insufficient molecules in the bulk of the fluid to allow the concentration to remain constant as the area is changed. With such molecules, known as *surfactants*, the concentration of the substance in the surface will be reduced as the area is increased. The surface tension will thus decrease as the area of surface is increased.

The lung contains a surfactant (a lipoprotein). Thus, as the alveoli become smaller the surface area of the fluid around the walls becomes less, and so the surface tension drops. The pressure inside the small alveoli will thus be less than it would have been in the absence of the surfactant, and alveoli of different radii will thus have more closely equal pressures arising from the surface tension forces of the fluid. By this means it is possible for all sizes of alveoli to become filled with air. A fatal disease in newborn children (whose lungs will with air for the first time at birth, having previously been liquid-filled) is one in which this surfactant is absent. The abnormally high surface tension that results prevents the alveoli from filling the air. The disease is known as hyaline-membrane disease or atelectasis.

7.6   Capillary rise

A well-known phenomenon is the way in which liquids rise up narrow-bored tubes. This is known as capillary rise. Provided that the diameter of the tube is sufficiently small, then the surface of liquid in that tube will be spherical (if the tube is too broad then there will be a flat region between the curved surface at the edges, caused by the angle of contact between the liquid and the material of the tube). The capillary rise occurs due to the pressure difference across the spherical surface. In Fig. 7.5 the pressure at D will be greater than that at C by that amount necessary to maintain the spherical surface. If the radius of the tube is r, and the angle of contact between the water and the surface of the tube is θ, then the radius of curvature of the water surface will be $r/\cos\theta$. From equation 7.4 the pressure difference is thus $2T\cos\theta/r$. The pressure at D must be atmospheric pressure, as must that at A, and thus at B. The pressure difference between points B and C is responsible for the rise of water in the tube. The pressure difference due to the surface tension will maintain a head of water giving this pressure. The pressure of a height h of a water column is $\rho g h$, where ρ is the density of water and g the acceleration due to gravity.

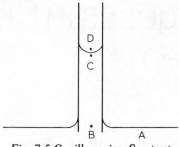

Fig. 7.5 Capillary rise. See text.

i.e.

$$2\,T\cos\theta/r \;=\; \rho g h$$

$$h \;=\; 2\,T\cos\theta/\rho g r$$

7.7 Questions

1. The xylem of many trees has a diameter of about 50 μm. What rise of water up the xylem would you expect due to capillary action, if the angle of contact were (a) 0° (b) 60°?

2. A clean glass tube of internal diameter 50 μm is immersed in water. How far up the tube will the water rise? The tube is now immersed until only 1 cm is above the surface. What happens?

3. How much work, against surface tension forces, is required to pull a stick 2 cm long and 4 mm in diameter vertically through a water surface?

8 Energetics in Biological Systems

An essential feature of life is that its maintenance requires the continual input of energy. This chapter and Chapter 9 are concerned with the energetics of an organism at several different levels; from the energy balance of the whole organism — the different forms of input and output of energy with respect to the entire organism — to the energetics of the individual chemical reactions within the cell.

8.1 Conservation of energy — the first law of thermodynamics

One underlying principle of energetics, first mentioned in Chapter 4, is the conservation of energy. Thus for any system we can keep an 'energy account' — any change in the total energy of the system is equal to the difference between the energy entering and the energy leaving the system. We use the word 'system' to mean the unit under discussion — this for example may be the whole organism, or just a part of a cell (the same laws apply in all cases).

Energy can be gained and lost by a system in three ways:

1. By transfer of heat.
2. By performance of work.
3. By transfer of matter.

In different types of system these three contributions can have very different importance. Consider the case of the ecologist wanting to determine the energy balance of an animal. Heat transfer is occuring continuously, since the animal is generating heat due to its metabolic activity, and is receiving heat from its surroundings. It will reach an equilibrium temperature when the net rate of heat loss from the body is equal to the rate of metabolic heat production. The problem of heat transfer is discussed in Chapter 9 and will not be considered further here. Likewise there is a continual transfer of matter between the animal and its surroundings. The animal feeds and takes in a variety of foodstuffs and water, and it respires, taking in oxygen. It also loses matter in a variety of forms, including faeces, urine, exhaled gases, sweat and so on. However, at first sight maybe surprisingly, most animals (apart from man and his draft animals such as horses, oxen etc.) do very little net work on the surroundings, when averaged out over a period of time. In the short term they do work in the form of muscular movements, but virtually all the energy used by the animal to perform these movements eventually leaves the animal in the form of heat. There are of course many situations in which external work is performed – a burrowing animal lifts soil to the surface for example, but the amount of work involved is insignificant compared with the other forms of energy transfer.

On the other hand consider the muscle physiologist wanting to work out an energy balance for his muscle during a single twitch. In this case, since the time interval is so short, there will be very little transfer of matter to and from the entire muscle cell. There will on the other hand be both heat transfer and the performance of work. We can see that the above three contributions to the energy account can have relatively different degrees of importance in different situations.

Let us consider the energy content of matter a little more closely. The total energy content will be distributed in many different forms – from the energy contained within the nuclei due to the nuclear forces, to the energies of the various electrons of the molecules in their energy levels, to the energies of rotation and vibration of the molecules. There will be many other forms of energy contained within the matter being considered, including a contribution from the way in which the total energy is distributed amongst the various energy levels, to be discussed later and known as entropic or bound energy. As biologists we are not usually interested in the total energy content of the matter, we are more concerned with the energy *change* that occurs during some reaction. Consider the situation for the moment in which there is no transfer of matter, so that the only way in which energy can be exchanged between the system containing the matter and the surroundings is by heat flow or by the performance of work. Any change of energy which does take place is said to change the *internal energy*, U, of the system. The internal energy of a system is a concept similar in type to that of temperature, and is a collective term for all the separate contributions mentioned above (plus all those not mentioned!). To say that a system has so much internal energy says nothing at all about the way the energy is distributed. For a closed system (one in which there is no transfer of matter) we can write

$$\Delta U = \Delta Q + \Delta W \tag{8.1}$$

where the symbol Δ means 'the change in' and ΔQ and ΔW are the net amounts of heat and work entering the system. For the general situation in which transfer of matter is possible (an open system):

$$\Delta U_{sys} = \Delta Q + \Delta W + \Delta U_m \tag{8.2}$$

in which ΔU_{sys} stands for the change in internal energy of the system and ΔU_m denotes the net increase of internal energy of matter entering and leaving the system.

Work can be performed either on or by the system in several different ways. If there are changes in volume, then work will be done equal to PV; if there are length changes (as for example in muscle) then work will be done equal to $T_n L$ (see page 57 for fuller explanations). Likewise if the electrical potential of a system of net charge is changed, or if the concentration of a solution is altered. Likewise work is required to increase the surface of a material exhibiting surface tension (as in the lungs), to move a charged particle down an electrical potential gradient (for example across a membrane in nerve or muscle) or to move a chemical along a concentration gradient in a solution.

Since biologists are normally concerned with systems operating at a constant pressure we can automatically include the effect of volume changes taking place at this pressure by defining a new variable, known as the *enthalpy* of *heat content* (H).

$$H = U + PV \qquad\qquad (8.3)$$

where P and V are the pressure and volume of the system. Since the pressure is constant.

$$\Delta H = \Delta U + P \Delta V \qquad\qquad (8.4)$$

We can illustrate the importance of enthalpy to a biologist by considering the combustion of, say, stearic acid to carbon dioxide and water.

$$C_{18}H_{36}O_2(s) + 26O_2(g) \longrightarrow 18CO_2(g) + 18H_2O(l) \qquad\qquad (8.5)$$

(The s, g, l in parentheses denote that the substance is in solid, gaseous or liquid form).

If the combustion takes place at a constant volume, then no external work is done on the system, and from equation (8.1) above we see that the heat lost by the system if it is kept at a constant temperature is just equal to the change in the internal energy.

However, if the combustion takes place at a constant pressure (as happens in the body), then external work is done, since the products contain a different amount of gas (18 moles/mole of stearic acid) to the reactants (26 moles/mole of stearic acid). The heat lost by the system in this case will be

$$\Delta Q = \Delta U + P \Delta V \qquad\qquad (8.6)$$

Note that we have written $+ P \Delta V$ rather than $- P \Delta V$, since we are here considering the work as work done *by* the system. When writing equations 8.1 and 8.2 we considered ΔW as the work done *on* the system. Thus in this case, doing the experiment at constant pressure, the heat change measures the enthalpy change, not the change in internal energy. Of course we can immediately convert one to the other, since we can easily determine the value of the work done, $P \Delta V$.

The heat of combustion of stearic acid, measured in a bomb calorimeter (i.e. at constant volume) is about $- 11\,380$ kJ/mole. This is thus a measure of the internal energy change U. If measured at constant pressure, then the volume change that occured would have been due to the change in the net amount of gas. In this case, using the perfect gas equation

$$P \Delta V = \Delta n\, RT = -8 \times 8.4 \times 300/1000 \text{ kJ/mole}$$

$$\doteq -20 \text{ kJ/mole} \tag{8.7}$$

Thus the change in enthalpy is $-11\,360$ kJ/mole. The difference is extremely small, but because we are working at a constant pressure we talk in terms of enthalpy changes rather than internal energy changes.

Thus when we are considering energy-accounts under conditions when transfer of material is an important factor we can use

$$H_{sys} = H_i - H_o + Q_i - Q_o + W_i - W_o \tag{8.8}$$

in which H_{sys} is the change in the enthalpy of the system, H_i and H_o are the enthalpies of the matter entering (i) and leaving (o) the system, Q_i and Q_o are the heat transfer and W_i and W_o denote the work done on and by the system *other than that necessary to keep the system at a constant pressure*.

8.2   Energy conversion

Another feature of energetics is that one form of energy may be converted to other forms. This is a common feature of everyday life. Thus the chemical energy of petrol is converted into mechanical energy in the internal combustion engine. Mechanical energy is converted into electrical energy in the dynamo. Chemical energy can be converted directly into electrical energy in the battery or fuel cell. The list is endless. Likewise in biological systems mechanisms exist for converting chemical energy into mechanical energy (muscles, cilia), chemical energy into light (luminescent organs), light into chemical energy (photosynthesis) and so on.

A general feature of such conversion is that some of the energy being converted is always lost in the form of thermal energy. Further, whereas man-made machines can be made which convert heat into other forms of energy, these machines all require differences of temperature between different parts of the system for their operation. Living organisms have no such capability and thus can make no use of thermal energy for conversion to other forms of energy. Thus the generation of heat results in energy wasted as far as the organism is concerned.

8.3   Free energy and chemical potential

We have so far discussed the total energy content of a system; we now need to consider the useful energy, meaning that fraction of the total energy content that can be used to perform useful work in one form or another. We shall see later that not all of the total energy is available for such work.

As biologists we are most interested in the amount of work that can be obtained under conditions in which the number of molecules of a particular chemical substance changes. This includes for example the situation in which a chemical reaction proceeds, so that a certain number of reacting molecules are transformed to become a certain number of product molecules. Here we have a reduction in the number of reactant molecules and an increase in the number of product molecules. How much work is involved in removing or adding molecules to a solution already containing a certain concentration of those molecules? It includes the situation in which a membrane, permeable to some molecules only, separates solutions of different substances. How much work is involved in transporting a molecule of a particular kind from one side of the membrane to the other? These molecules may be charged. What difference does this make? The change in the number of molecules in this case is a decrease in the number of one species on one side of the membrane, and an increase on the other. As a final example it includes the situation in which water molecules are transported from the bottom of a tree to the top, and evaporate through the leaves into the external environment. Here we are concerned with the amount of work done when molecules leave a container with a certain amount of rigidity, so that the loss of molecules of water results in a reduction of pressure — in other words, how much work is done in transporting molecules against a pressure gradient?

The general term describing the change in *free energy* (G) (meaning that energy available for useful work) as the number of molecules of a particular species enters or leaves the system under consideration is known as the *chemical potential*. This term would perhaps be better understood if it were known as *chemical potential energy*. It is exactly analogous to the mechanical potential energy that is a more familiar concept. Mechanical potential energy is the energy that a body has due to its position. Thus a body of mass m is said to have:

$$\text{Mechanical potential energy} = \alpha_0 + mgh \qquad (8.9)$$

where h is the height of the body above some reference position, and g is a constant (the acceleration due to gravity). At the reference position the mechanical potential energy is some arbitrary value α_0.

There are three contributions to the chemical potential that we need to consider:

1. The energy due to the composition or concentration of a molecular species.
2. The energy contribution for charged ions due to the electrical potential energy of the solution.
3. The contribution due to the pressure of the system.

We shall not consider the contribution due to pressure until the end of this chapter. Until then we shall consider the more common biological situation, in which we have a constant pressure and temperature, and we shall obtain expressions for the first two contributions to the chemical potential.

Energy of a substance due to its concentration

That a solution (or a mixture of gases) should have energy associated with the concentration of a particular substance is straightforward, and is due to the fact that, in the absence of any restraint, the substance will diffuse to be at a lower concentration. The expression relating the chemical potential to the concentration is:

$$\text{Chemical potential (energy) per mole} = \mu = \mu_0 + RT \ln (\text{concentration}) \quad (8.10)$$

In this expression R is the gas content (with units of energy/mole/$^\circ$K) and T is the absolute temperature. ln means 'log$_e$'. Thus the chemical potential defined like this has units of energy/mole. The *total* energy of n g-mole of a substance whose concentration in a solution is C g-mole/litre is

$$n\mu = n\mu_0 + nRT \ln C \quad (8.11)$$

The units are now of energy. Once again we have had to refer to a reference level (μ_0) which is taken to be the chemical potential of a molar solution, and is known as the *standard chemical potential*.

This formula has been stated without any derivation. Those interested in the derivation should attempt question 3 at the end of the chapter. A full solution is given at the end of the book for those who have difficulty.

Energy of a substance due to its charge

The electrical potential energy of a system is the energy that a system has due to its *charge*. It arises because of the fact that like charges repel one another, and hence energy is required (we must do work) to bring like charges into the same environment. They have a repulsive force and hence tend to disperse. Potential differences (measured in volts) are a familiar concept. These are a measure of the work done in taking unit charge (one coulomb) from one level of potential to another.

$$\text{Electrical potential energy} = E_0 + E_m \quad (8.12)$$

with units of energy/coulomb (called volts). E_0 is an arbitrary reference level. This is always taken to be the potential of the *earth* and set at zero. If we want to find the potential energy of one mole of charge we must multiply the potential as defined above by the number of coulombs per mole. For a univalent ion this is the Faraday (F). For an ion of valency z, the charge/mole is zF. If we have n g-mole of this ion in our system the total electrical potential energy is

$$\text{Electrical P.E.} = nzF E_m \quad (8.13)$$

Our units are now those of energy once again.

The normal type of situation which we encounter is one which involves a *change* from one set of circumstances to a second set of circumstances. We are interested in the energy changes which take place during this change. As biologists there are two particular situations which we must understand. The first involves chemical reactions, when we wish to follow the energy changes taking place as a set of reactants are converted to a set of products, and we want to be able to determine the extent to which the reaction will occur. The second involves the transport of substances across membranes when we want to be able to understand the energetics of such transport, and the way in which equilibrium is established.

The first thing that we must consider is the magnitude of the energy changes involved when we go from one state to another. In terms of our two examples this means 'How much energy is released or absorbed when our reactants are converted to our products?' and 'How much energy is required to transport a substance across a membrane?'

Asking ourselves 'how much energy is required' is the same as saying 'How much work must we do'. In the case of our mechanical potential energy, the work we have to do to lift the body from one height (h_1) to another (h_2) is simply given by the *difference of potential energy* of the two positions.

$$\text{Work} \; = \; \text{difference of P.E.} \; = \; mg(h_2 - h_1) \tag{8.14}$$

In the same way we find that the work done in transferring 1 mole of a chemical from a solution in which the concentration is C_1 to one in which the concentration is C_2 is simply given by the *difference of chemical potential* between the two solutions. That is

$$\text{Work} \; = \; \mu_1 - \mu_2 \; = \; RT \ln C_2 - RT \ln C_1 \; = \; RT \ln (C_2/C_1) \tag{8.15}$$

For n moles the work is n times as great. Notice that in both these examples, when we have considered changes in the system, the reference levels have cancelled out. This is not always the case, particularly when we consider chemical reactions.

We can also evaluate the work done in transferring charge from one potential to another. For n g-moles of an ion of valency z (whose total charge therefore is nzF)

$$\text{Work} \; = \; nzF \, (E_2 - E_1) \tag{8.16}$$

We can understand now why we talk about the free energy of a system as being the work we can get our of the system. This is because the work obtained from a change in the system is given by the difference between the initial and final values of the free energy; that is

$$\text{Work} \; = \; \text{final free energy} - \text{initial free energy} \tag{8.17}$$

We can sum these two contributions to the chemical potential to provide a single expression:

$$\text{Chemical potential } \mu = \mu_0 + RT \ln (\text{concentration}) + zFE \tag{8.18}$$

8.4 Entropy

Not all the energy within a system is free energy. The remainder is known as the entropic energy, and is the product of the *entropy* of the system and the absolute temperature. Thus the total energy of the system, which is the enthalpy, and which we have already subdivided in one manner (into internal and external energies), can be subdivided in another way into *free* and bound or entropic energy

$$H = U + PV = G + TS \qquad (8.19)$$

The easiest concept of entropy is one of order. Any system will have a tendency to become disordered. The technical way of saying this is to say that the system tends to increase its entropy. Energy is required to restore order to the system. An example of this is the mixing of two solutions of different solutes. Initially these are quite distinct, and the system is ordered (has low entropy). After mixing the system is disordered (has high entropy). To 'unmix' the solutions requires energy. Thus the state of order of disorder of the system (the magnitude of the entropy) contributes to the total energy of the system. We find that the energy associated with the state of entropy is also dependent upon the temperature since the thermal energy of the molecules in the system increases with temperature thereby increasing the ease with which the system can become disordered.

Why do we make a distinction between the *free energy* and the *entropic energy* when talking about the total energy of a system? The answer becomes apparent when we consider the equilibrium of the system.

8.5 Equilibrium

In the mechanical system that we have considered the body will fall until it has reached a position of minimum potential energy. Likewise our biological systems will change until they have reached a situation in which their energy is a minimum. The question is — which one of the many energies that we have been considering? The interesting answer is that it is not the total energy (the enthalpy) but the free energy which is minimized. We can understand this by considering the chemical reaction

$$A + B \rightleftharpoons C + D \qquad (8.20)$$

As this reaction proceeds there will be some change in enthalpy, since the heat content of the products will be different from that of the reactants. There will also be a change in the entropy, since there will be some difference in the degree of order of the products as compared with that of the reactants. There is nothing that can be done about this. The process of reacting necessarily leads to the change in the order, or entropy, of the system. Thus even though the heat content is changing (the system is liberating or absorbing heat from the surroundings) so is the energy bound up in the entropy of the system. We can speak of the enthalpy change as being a tendency to cool, and the entropic energy change as being the tendency to disorder. The net energy change in the reaction is the difference between these, and this difference is known as the free energy

$$H - TS = G \tag{8.21}$$

This is the same equation as we met above. The final equilibrium position of the reaction is thus a balance between the enthalpy changes and the entropy changes in the system. Thus:

For equilibrium, the Free Energy (G) is minimized.

Let us consider the equilibrium reaction again. We can work out the total free energy of the system (free energy of reactants plus free energy of products) for any degree of completeness of the reaction. Suppose we plot a graph of free energy against the proportion of reactants and products present. This will have the form:

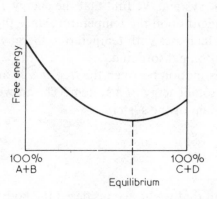

Fig. 8.1

Equilibrium will be reached at the minimum value of the free energy (G), that is when

$$dG = 0 \tag{8.22}$$

This is an alternative way of defining equilibrium.

In this reaction, the only quantities which are changing are the concentrations of the different substances involved. Therefore the only contribution to the free energy which is changing is the chemical potential. The usefulness of the chemical potential is thus apparent, since under conditions in which *only molecular concentrations* are involved

Equilibrium is obtained when $d\mu = 0$ \hfill (8.23)

where $d\mu$ in this case is the total change in chemical potential of all the substances involved.

We shall now look at (a) chemical reactions and (b) movements across membranes in more detail.

8.6 Chemical reactions

Once again let us consider the reaction

$$A + B \rightleftharpoons C + D \tag{8.20}$$

We can write down the work that is done (difference in free energy ΔG) when the reaction proceeds simply by subtracting the free energy of the products from the free energy of the reactants. Since only the chemical concentrations are changing this difference in free energy is given by the difference in chemical potential.

Thus work done $= \Delta G = \Delta\mu = (\mu_C + \mu_D) - (\mu_A + \mu_B)$

$$\Delta\mu = (\mu_{0C} + RT \ln C + \mu_{0D} + RT \ln D) - (\mu_{0A} + RT \ln A + \mu_{0B} + RT \ln B)$$

$$\Delta\mu = (\mu_{0C} + \mu_{0D}) - (\mu_{0A} + \mu_{0B}) + RT (\ln C + \ln D - \ln A - \ln B)$$

$$\Delta\mu = \Delta\mu_0 + RT \ln \frac{CD}{AB} \tag{8.24}$$

where $\Delta\mu_0$ is the difference of standard chemical potentials of reactants and products. A, B, C and D now refer to the concentrations of the substances. Under these conditions (only concentrations involved) we can write this as

$$\Delta G = \Delta G_0 + RT \ln \frac{CD}{AB} \tag{8.25}$$

These equations represent the work done when one g-mole of each of the reactants is converted to one g-mole of each of the products under conditions in which the *concentrations* of the products and the reactants are kept *constant*. If we convert n g-moles, then both sides of each of these equations must be multiplied by n.

i.e. for n g-mole

$$\text{Work done} = n \Delta G = n \Delta G_0 + nRT \ln \frac{CD}{AB} \tag{8.26}$$

This requirement, that the equations only give the work done when the concentrations remain constant, is very important. The effect of the reaction proceeding is of course to change the concentrations. It is quite straightforward, though tedious, to evaluate the work done as the reaction proceeds, by integrating the expression above over the change in concentrations. However, the restriction is not so bad as might at first sight appear. In the first place we can restrict ourselves to considering the work done when the reaction proceeds by such a small amount that there is no significant change in the concentrations. More importantly, from a cell's point of view, is the fact that reactions inside a cell do not usually occur in isolation. Normally there is a whole chain of reactions, so that as our particular reaction proceeds so the reactants are replenished and the products used as reactants for further reactions. For most of the most relevant reactions in the cell the concentrations of reactants and products remains remarkably constant. How the cell maintains this state of affairs is an interesting problem, but outside our scope here. Suffice it to say that the above expression for the change in free energy is usually relevant without modification to biological reactions.

Let us consider the equilibrium position of this reaction. We saw in the last section that equilibrium is obtained when $dG = 0$. Thus, for equilibrium

$$\Delta G = 0 = \Delta G_0 + RT \ln \frac{CD}{AB} \tag{8.27}$$

Since ΔG_0 is a constant, dependent only on the molecular species involved in the reaction it follows that

$$RT \ln \frac{CD}{AB} \text{ is constant}$$

which is only true if

$$\frac{CD}{AB} = \text{constant } K \tag{8.28}$$

K is known as the *equilibrium constant* for the reaction. Working backwards from these equations we see that

$$\Delta G_0 = -RT \ln K \tag{8.29}$$

What is the meaning of ΔG_0? This also becomes clear now. If the concentrations of both products and reactants are maintained equal (i.e. $A = B = C = D$) then the free energy change obtained by causing 1 mole of each of the products to be formed from each of the reactants is equal to ΔG_0 which is known as the *standard free energy* change of the reaction.

There are two important aspects of this from a biochemist's point of view:

1. By knowing the equilibrium constant he can tell the extend to which the reaction will proceed. This is only true provided the rates at which reactants are supplied and products are removed are slow compared with the rate of the reaction itself. For many reactions in the cell this is true.

2. He can determine the energy changes occurring as the reaction proceeds. Let us illustrate this. One very important reaction in the cell is the hydrolysis of ATP

$$ATP + H_2O \rightleftharpoons ADP + H_3PO_4 \tag{8.30}$$

This reaction is used to provide the energy to drive many processes taking place in the cell. The standard free energy change (ΔG_0) is approximately -7000 cal/mode. This means that 7000 calories are released when one mole of ATP is split *under conditions in which the concentrations of ATP ADP and P_i are all maintained at 1 molar*. This, except perhaps for a rather bizarre test-tube experiment, is never the case. The actual amount of free energy change in the cell is dependent upon the concentrations of the molecules in the reaction, and will be different for different concentrations. Thus when asked the question 'how much energy is released by the hydrolysis of ATP?' you must first determine what these concentrations are. For example the concentrations of these substances in a muscle cell are approximately:

ATP – 0·01 M
ADP – 0·0008 M
P_i – 0·015 M
H_2O – This is taken by convention to be unity.

With these concentrations the free energy of hydrolysis of ATP is

$$\Delta G = \Delta G_0 + RT \ln \frac{0\cdot015 \times 0\cdot0008}{0\cdot010 \times 1} = 7000 + 600 \ln (1\cdot2 \times 10^{-3})$$

$$= 7000 + 600 \times 2\cdot3 \times (-2\cdot92) \doteq -11\,000 \text{ cal/mole}$$

Under these conditions (which are approximately those occurring in many cells in the animal) the energy available for use by the cell is considerably greater than the standard free energy. Under natural conditions the cell has other reactions which keep these concentrations approximately constant. Let us consider what would happen however if for some reason the concentration of ADP in the cell rose from 0·8 mM to 8 mM (i.e. from 0·0008 to 0·008 M). The free energy released by the hydrolysis is now

$$\Delta G = -7000 + 600 \ln (1\cdot2 \times 10^{-2}) = -7000 + 600 \times 2\cdot3 \times (-1\cdot92)$$

$$-9650 \text{ cal/mole}$$

Thus the available energy has been considerably reduced.

We are also in a situation, knowing the standard free energy change, to evaluate the equilibrium constant for the reaction. We know that

$$\Delta G_0 = -RT \ln K$$

$$\text{i.e. } K = \exp/(-\Delta G_0/RT)$$

In this case therefore $K = \exp(7000/600) = \exp(11\cdot7) = 120\,000$. Since $K = $ ADP $\cdot$ P_i/ADP $\cdot$ H_2O this means that under conditions of equilibrium there is considerably more ADP than ATP.

Normally, in the cell, conditions are very far from equilibrium. How can this be the case — why isn't ATP being continually hydrolysed by the cell? How can one make up solutions of ATP for biochemical experiments? These difficulties are overcome when we come to consider the *rate* at which reactions occur. So far we have only discussed the *equilibrium* conditions of a chemical reaction. This distinction between the final equilibrium conditions and the rate with which those conditions are attained is extremely important, and for an understanding of what is actually going on in the cell we need to know about both effects. We have seen that *free energy*

considerations determine equilibrium. The concept of *activation energy* determines
the rates of reactions. We shall not discuss activation energies here. They are well
explained, for example, in Wallwork 'Physical Chemistry'. However, the reason why
ATP is not continuously hydrolysed at a high rate is because the activation energy
for its hydrolysis is large. (In fact, in a solution, the ATP is continuously hydrolysed
at a low rate; for this reason it is normally stored at as low a temperature as possible).
The purpose of an enzyme is then to catalyse the reaction (which means reducing
the activation energy). There are many enzymes which cause the hydrolysis of ATP.

8.7   Membrane phenomena

We shall illustrate the factors which must be taken into account to understand the
phenomena which occur at membranes by considering the way in which *membrane
potentials* arise.

   Let us consider what will happen if a membrane which is permeable *only* to
potassium ions separates unequal concentrations of the potassium salt of a large
organic anion (A^-).

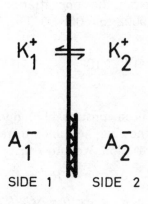

Fig. 8.2

In this case potassium ions will initially diffuse more rapidly into the side with the
lower concentration than into the side with the higher concentration. This will cause
there to be an imbalance of charge on either side of the membrane, which will result
in an electrical potential difference being set up across the membrane. This will cause
the rate at which ions diffuse into the side with the lower concentration to slow down,
and equilibrium will be reached when the flow in one direction just equals the flow
in the opposite direction.

   Equilibrium will be reached when the free energy of the potassium ions on one
side equals the free energy of the potassium ions on the other side. In this situation

it is only the free energies of ions which are free to cross the membrane (ions to which the membrane is permeable) which will balance out at equilibrium. Thus once again we have that the condition of equilibrium is:

Difference of free energy of diffusable ions is zero.

We have derived expressions for the free energy of a solution of ions. Thus at equilibrium

$$\Delta G = \Delta \{n(\mu + zFE)\} = 0$$

$$\text{i.e. } \mu_1 + zFE_1 - \mu_2 - zFE_2 = 0$$

$$RT \ln K_1^+ - RT \ln K_2^+ = zF(E_2 - E_1) = zF\Delta E_m$$

where ΔE_m is the potential difference across the membrane.

Hence
$$\Delta E_m = \frac{RT}{zF} (\ln K_1^+ - \ln K_2^+)$$

$$\Delta E_m = \frac{RT}{zF} \ln (K_1^+/K_2^+) \tag{8.31}$$

This is known as the Nernst equation. At room temperature inserting the values of the constants we find

$$\Delta E_m = 58 \log_{10} (K_1^+/K_2^+) \, mV \tag{8.32}$$

Since the valency of potassium is 1, this value was used for z in the Nernst equation. Has we been considering a divalent ion (e.g. Ca^{++}) then we should have used $z = 2$. In this case

$$\Delta E_m = 29 \log_{10} (Ca_1^{++}/Ca_2^{++}) \, mV \tag{8.33}$$

The problem which confronts us is to know what the concentrations K_1^+ and K_2^+ are. How many K^+ have to cross the membrane to give rise to the potential difference ΔE_m. This is determined by the membrane capacitance. It is found that with normal membranes (which have a capacitance of about $1\mu Farad/cm^2$) the number of ions which have to be transported across is so few as to make no detectable change in the ionic concentrations on either side. Hence we can use our starting concentrations in the formula to determine the membrane potential.

For example, in nerve fibres the external concentration of K^+ is about 10 mM, whereas the internal concentration is about 400 mM. Thus, if the membrane were permeable *only* to potassium ions the membrane potential would be

$$E_K = 58 \log_{10}(10/400) = -93 \, mV \tag{8.34}$$

We have written E_K rather than ΔE_m to show that we are determining the potential the membrane would have if it were permeable only to K^+. This is known as the *potassium equilibrium potential*. (In a real nerve the membrane is permeable to other ions also, to a lesser extent. The consequence of this is that the membrane potential is smaller than the potassium equilibrium potential).

The example we have given above, with the membrane permeable only to potassium ions, is a particular case of what is known as the *Donnan equilibrium*. A more general illustration of this is provided by the example shown below.

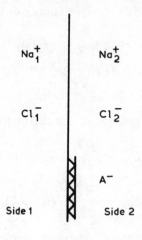

$$Na^+_1 \qquad\qquad Na^+_2$$

$$Cl^-_1 \qquad\qquad Cl^-_2$$

$$A^-$$

Side 1   Side 2

Fig. 8.3

Here the membrane is permeable to all ions except large organic anions, labelled A^-. Under these conditions equilibrium will be set up for all the ions to which the membrane is permeable, in the same way as before. Thus we find that for the solium ions.

$$E_m = \frac{RT}{F} \ln \frac{Na_1}{Na_2} \tag{8.35}$$

and likewise for the chloride ions

$$E_m = \frac{RT}{F} \ln \frac{Cl_2}{Cl_1} \tag{8.36}$$

Notice that in these two equations the suffices are inverted. This is because of the different sign of the charge of sodium compared with that of chloride. Since the membrane potential has a unique value it follows that

$$\frac{RT}{F} \ln \frac{Na_1}{Na_2} = \frac{RT}{F} \ln \frac{Cl_2}{Cl_1}$$

which is only true if

$$\frac{Na_1}{Na_2} = \frac{Cl_2}{Cl_1} \tag{8.37}$$

This is the condition for equilibrium in the system, and determines the magnitude of the membrane potential in terms of the concentrations of sodium and chloride on each side of the membrane. We have not so far evaluated the magnitude of these concentrations. In order to do so we need to know:

1. The total amount of sodium, chloride and anion in the system.
2. That there is electrical neutrality on each side of the membrane.

Both these facts are measured in terms of numbers of molecules, whereas equation 8.37 was using concentrations. If we make the simplification that the volume on each side of the membrane is 1 litre, then we may use concentrations to express points 1 and 2 giving:

$$Na_1 + Na_2 = Na_t \tag{8.38}$$

$$Cl_1 + Cl_2 = Cl_t \tag{8.39}$$

$$Na_1 = Cl_1 \tag{8.40}$$

$$Na_2 = Cl_2 + A \tag{8.41}$$

The equations 8.37–8.41 completely specify the system, and they can now be manipulated to determine the unknown quantities (Na_1, Na_2, Cl_1 and Cl_2) in terms of the known quantities (Na_t, Cl_t and A). You will find that

$$Na_1 = Cl_1 = \frac{Na_t(Na_t - A)}{2Na_t - A} \tag{8.42}$$

$$Na_2 = \frac{Na_t^2}{2Na_t - A} \tag{8.43}$$

$$Cl_2 = \frac{(Na_t - A)^2}{2Na_t - A} \tag{8.44}$$

It is *not* important that you remember these three equations, you are unlikely to use them. It is important that you understand point 2. above, and the electrochemical potential condition of equilibrium. These are at first sight inconsistent, since the development of a potential difference across the membrane depends upon there being a slight imbalance of charge. Once again let us make the point that with normal membranes the difference of charge required to give the measured potentials is so small that the divergence from electrical neutrality on each side of the membrane is completely negligible.

8.8   Contribution of pressure changes to chemical potential

So far we have been concerned in this chapter with the contribution to the chemical potential energy of a system by the concentration and electrical potential. Equation

8.18 gives the magnitude of μ in terms of these two contributions

$$\mu = \mu_0 + RT \ln(\text{concentration}) + zFE \qquad (8.18)$$

Suppose now however that we remove molecules of a particular species from one container to another. If the pressures of the two containers are kept constant and equal, then there will be a volume decrease in the container from which the molecules are removed, and an equal volume increase in the other container. The magnitude of this volume change is given by the quantity known as the *partial molar volume* ($\bar{V}$) times the number of moles of that species moved. The partial molar volume is discussed further in question 4 at the end of the chapter. It is of the same order of magnitude as the volume of one mole of the substance, but differs due to the effect on the water molecules of the substance when it is dissolved. Thus the change in volume is $\bar{V} \, dn$ where dn is the number of moles of substance transferred. The work done in changing the volume of the solution at a pressure P is given by the product of the pressure and the change in volume.

$$\text{Work} = P \bar{V} \, dn \qquad (8.45)$$

Provided the pressures of the two containers are equal, then the work done in removing the molecules from the one container is exactly balanced by the work done in adding the molecules to the other container (assuming that the partial molar volumes are the same − this will be true only provided the solutions are dilute). However, if the first container is at a pressure dP greater than that of the second container, then an extra amount of work will be involved, and this extra, net work will be given by

$$\text{Net work} = dP \, \bar{V} \, dn \qquad (8.46)$$

However, in terms of the chemical potentials of the two solutions the net work equals $dn \, d\mu$

$$\text{i.e. } dn \, d\mu = dP \, \bar{V} \, dn$$

$$d\mu = dP \, \bar{V} \qquad (8.49)$$

The contribution to the chemical potential of a solution of a particular species is thus given by

$$\mu = P \bar{V} \qquad (8.47)$$

and our equation for the chemical potential of a solution due to the three contributions (concentration, electrical potential and pressure) is:

$$\mu = \mu_0 + RT \ln(\text{concentration}) + zFE + P \bar{V} \qquad (8.48)$$

We need to be a bit careful about the meaning given to μ_0; this is the chemical potential of a reference solution, which is taken to be a one molar solution at zero volts and one atmosphere pressure.

8.9 Chemical potential of water

All the discussion so far has concerned the chemical potential of substances dissolved in solvents — for our purposes generally water — and it is not perhaps immediately apparent what we mean when we speak of the chemical potential of that solvent. In order to determine this we need to consider the phenomenon of *osmotic pressure*.

A common situation in biology is to have a membrane which is selectively permeable to specific molecules and impermeable to others separating two aqueous solutions. The simplest such situation we can envisage is one in which a membrane permeable only to water separates pure water on one side of the membrane from a solution of some substance on the other. In this situation there will be a tendency for the water to diffuse across the membrane from the side containing pure water to the side containing the solution. The tendency for there to be a net water transfer can be opposed by applying a positive pressure to the solution. It is found that for dilute solutions the pressure π required to obtain no net water transfer is given by

$$\pi = RTc \qquad (8.50)$$

where R is the gas content, T the absolute temperature and c the concentration in moles/litre. π is known as the osmotic pressure of the solution.

If two solutions have different osmotic pressures π_1 and π_2, then the work done in transferring dn moles of water from one solution to the other is

$$\text{Work for water transfer} = (\pi_1 - \pi_2)\,\bar{V}_w\,dn \qquad (8.51)$$

in which $\bar{V}_w$ is the partial molar volume of water in the solutions (see question 5) which is approximately equal to the molar volume of water (i.e. the volume occupied by 18 g of water — 18 cm^3). Now the work done in transferring the water must equal the difference in chemical potential of the water between the two solutions (since this is how we define the chemical potential). Thus

$$\mu_{w2} - \mu_{w1} = (\pi_1 - \pi_2)\,\bar{V}_w \qquad (8.52)$$

Notice that instead of defining the chemical potential of the water in terms of its concentration we have done so in terms of the osmotic pressure which is a function of the concentration of the solutes in the solution.

There will also be a contribution to the chemical potential of the water from the pressure of the solution, exactly as was derived in the previous section. Thus the total expression for the chemical potential of the water (since there is no contribution from the electrical potential — water being essentially uncharged) is:

$$\text{chemical potential of water } \mu_w = \mu_{w0} - \pi\bar{V}_w - P\bar{V}_w \qquad (8.53)$$

One further relationship is worth noting. If we have a solution and from this solution remove solute molecules so as to change the concentration, then the chemical potential of both the solute and of the water is changed. If the change occurs at constant pressure then

$$n_s d\mu_s + n_w d\mu_w = 0 \qquad (8.54)$$

where $d\mu_s$ and $d\mu_w$ are the changes in the chemical potential of the solute water and n_s and n_w are the numbers of molecules of solute and water present. This is known as the Gibbs–Duhem equation.

We can rewrite equation 8.53 as

$$\psi = \frac{\mu_w - \mu_{w0}}{\overline{V}_w} = P - \pi \qquad (8.55)$$

The expression on the left-hand side of this equation is known as the *water potential*. Since μ has units of energy/mole, and $\overline{V}_w$ has units of volume/mole, it follows that the units of ψ are of energy/unit volume, or joules/metre^3 = N m/m^3 = N/m^2 which are the units of pressure. (Of course this is as it should be since both the terms on the right hand side of the expression are pressures.) The water potential of a solution is a measure of the tendency for that solution to take up water, and has been given a number of alternative names such as suction potential, diffusion pressure deficit. Botanists often refer to π as the osmotic pressure, and P as the turgor pressure.

Notice the opposite sign of P and π is the above equations. This is because the higher the osmotic pressure the greater the tendency for water to flow into the solution, but the greater the mechanical pressure applied, the greater the tendency for water to flow out of the solution. Thus water will tend to flow into a cell with a greater negative mechanical pressure. It is important to understand this for the next section.

8.10   Why does water rise up tall trees?

The physics of the barometer

Suppose you make a water barometer. This can be done by taking a closed glass tube at least about 12 m long and filling it with water. Keeping it filled, immerse it in water, vertically, with the closed end uppermost. Slowly draw the tube out of the water. The tube will remain filled with water for quite a while. The pressure at the level of the water surrounding the tube will be atmospheric. The pressure at the top of the tube will become less and less as the tube is raised by an amount $\rho g h$, where ρ is the density of water, g the acceleration due to gravity and h the height of the top of the tube above the surface. When the pressure at the top of the tube equals the saturated water vapour pressure at the temperature of the experiment (about 17·5 mmHg at 20°C), then *cavitation* will occur. Cavitation is the process whereby bubbles of vapour are formed. Another example of cavitation is the bubbles of water vapour in boiling water. It occurs whenever, in bulk water, the saturated vapour pressure of the water equals that of the pressure on the water. Thus when the barometer tube is lifted sufficiently far out of the water, water vapour will form in a space above the water, whose level will then remain constant whilst the tube is lifted further.

Mercury-in-glass barometers are better than water barometers for two reasons. First, since the density of mercury is about 13·6 times that of water, the height of the tube necessary to measure the pressure is only about 1/13·6 as great, or about 760 mm. Secondly, the saturated vapour pressure of mercury is much less than that of water at normal temperatures (about 10^{-3} mmHg, rather than 17·5 mmHg), which means that the error in the reading, due to the fact that cavitation occurs at this pressure, will be much less.

Water in trees

The reason why water rises in trees is that water evaporates from the leaves from vessels (the xylem) which are sufficiently rigid that the loss of water results in the build up of a large reduction in mechanical pressure. From the foregoing it would appear that when the pressure reaches the saturated water vapour pressure, then cavitation will occur, with the creation of vapour. However, provided the vessels holding the water are sufficiently narrow in diameter, then cavitation does not occur until much lower pressures — in fact not until the pressure in the xylem is at least several tens of atmosphere *negative*. The reason for this must be that the water in the narrow tubes is always sufficiently close to the surface of the tubes that the inter-molecular forces are still different from those found in the centre of a large volume of water. The osmotic pressure in the xylem is negligible. Thus the water potential (or suction pressure) is very large and negative, due to the process of transpiration. If we measure the water potential in atmospheres, then the height that this water potential can raise the water above ground level is equal to the number of atmospheres below atmospheric pressure times the height (10·34 m) that water can be raised by one atmosphere. The height reached by tall trees can be in excess of 100 m, requiring therefore about 10 atmospheres of negative pressure.

The actual route of water loss in leaves seems not to be known with certainty, but is probably from the xylem of the veins, via the living parenchyma cells to the lacunae below the stomata. Little work seems to have been done on the mechanical strength of the various cells, but it is unlikely that the living parenchyma cells could withstand negative pressures of any magnitude. However, these cells have a large osmotic pressure, and thus the difference in water potential between the xylem and the parenchyma cells arises largely from the difference in osmotic pressure rather than mechanical pressure.

Transpiration is necessary for the raising of water up the tree. However, if transpiration stops, due to the leaves falling off in winter, or for any other reason, then the negative pressure will be maintained without cavitation provided that the xylem tubes are well sealed.

This theory for water transport in trees is often known as the *cohesion theory*, since

it depends upon the cohesive forces between water molecules. A crucial aspect of
the theory, which is too little emphasized in most textbooks dealing with this problem, is that it only works in narrow capillaries. In large tubes, say of the order of
1 cm in diameter or more, the limit to the height that water could be raised would
be about 10 metres, after which height cavitation would occur. Of course, if for
some reason in the narrow capillaries, air is introduced, then the action of that
capillary will stop. Note that the capillary rise of water in tubes of the size of the
xylem (about 50 μm) is about 30 cm only (see question 1 of Chapter 7). Capillary
rise from surface tension forces is totally inadequate to explain the way water rises
in trees therefore.

8.11   Questions

1. Muscle cells contain an enzyme (myokinase) which catalyses the following reaction

$$ADP + ADP \rightleftharpoons ATP + AMP$$

The equilibrium constant for this reaction is 2.27. Under resting conditions the
substances are present in the cell in the following concentrations ATP: 8 mM; ADP
0·5 mM; AMP unknown. What is (a) the concentration of AMP if equilibrium conditions prevail (b) the free energy change when 2 moles of ADP are converted to
ATP and AMP?
2. How much work is done by the sodium pump in nerve to pump out sodium under
conditions in which the internal sodium ion concentration is 10 mM, the external
sodium ion concentration is 250 mM and the membrane potential is -60 mV (inside
negative with respect to outside)? 1 J = 1 coulomb X volt. Can one ATP molecule
supply the energy to pump out one sodium molecule if the standard free energy change
for the reaction

$$ATP \rightleftharpoons ADP + P_i$$

at pH 7·0 is -7000 cal/mole, and the concentrations of ATP and ADP are 8 mM and
0·5 mM respectively and the concentration of inorganic phosphate (P_i) is 20 mM? (You
must convert concentrations to molar.)
3. The work done in expanding a gas is given by $P\,dV$ where P is the pressure and dV
is the change in volume. This is analogous to the work done in stretching an elastic
element, $T\,dL$, discussed on page 57. How much work is done in expanding a given
mass of gas from a pressure P_1 to a pressure P_2?
  For dilute solutions, Raoult's Law states that the lowering of vapour pressure of
the solvent is proportional to the number of solute molecules present. The work
done in changing the concentration of a given amount of solute must equal the work
done in changing the vapour pressure. Thus into the equation derived from the above
paragraph we can write $C = \text{const.}\ P$. This then gives the value for the work done in
changing the concentration of a solution, and thus the difference in chemical potential
between two solutions.

4. A 1% w/w solution of KCl is made by adding 1 g of KCl to 99 g water (so that the total mass of the solution is 100 g). A 1% solution has a density of $1 \cdot 0046 \times 10^{-3}$ kg m^{-3} and a 2% solution has a density of $1 \cdot 011 \times 10^3$ kg m^{-3}.

What change of volume occurs, per g-mole of added KCl when a 1% solution is converted to a 2% solution by the addition of KCl. This volume change is known as the *partial molar volume* of KCl at 1·5% concentration (the average of 1% and 2%). The molecular weight of KCl is 74. Compare the value you obtain with the molar volume of crystalline KCl, which is 37·5 cm^3.

The partial molar volume is concentration dependent. Determine the partial molar volume of 21% KCl, given that the densities of 20% and 22% KCl are $1 \cdot 1328 \times 10^3$ kg m^{-3} and $1 \cdot 1478 \times 10^3$ kg m^{-3} respectively.

5. The above data also enable us to determine the partial molar volume of water in these solutions. Thus we can determine the partial molar volume of a 1·5% solution by determining the volume change that occurs when a 2% solution is diluted to a 1% solution by the addition of 1 g-mole of water (18 g). Work out the partial molar volume of water in 21% KCl solution.

6. What is the osmotic pressure of sea water. Assume that this contains 500 mM NaCl.

9 Heat, Heat Flow and Diffusion

The idea of *temperature* is a familiar concept. This chapter is largely concerned with differences of temperature. What is the difference between an object at two different temperatures? How does heat 'flow' from one point to another? What are the important factors controlling the transfer of heat to and from objects, in particular biological objects (plants and animals).

We must start by asking what we mean by 'heat' and 'temperature' a little more closely, and for this we need to go down to the molecular level of matter for our understanding. The best way of understanding temperature and heat is in terms of the kinetic theory of gases. This is being included here in order to show how temperature is introduced into the theory. It is introduced as a basic postulate by assuming that *the mean kinetic energy of the molecules of the gas is proportional to the absolute temperature*. Thus the hotter the gas the greater the energy of the molecules. This idea does not only apply to gases but is true also for liquids and solids.

9.1 Kinetic theory of gases

There are a number of fundamental assumptions made about gases and which form the foundation of the kinetic theory. The success of this theory is its ability to explain in simple terms a whole host of physical phenomena, and it was instrumental in the development of ideas about the molecular nature of matter. These assumptions are:

1. All gases are made up of molecules, differing in size and mass for different gases.
2. These molecules are in constant, rapid and random motion, continually colliding with one another and with the walls of any container enclosing the gas (and thus constantly changing speed and direction).
3. The collisions are perfectly elastic. Thus there is no loss of kinetic energy in the collisions.
4. The molecules obey Newton's laws of motion (Chapter 4).
5. There are no forces of attraction of repulsion between the molecules, and the time

116

spent in collision is small compared to the time between collisions.

6. The temperature of the gas is proportional to the mean kinetic energy of the molecules.

Given these assumptions the kinetic theory applied Newton's Laws to the movements in order to derive an expression relating first the pressure of the gas to the movement, and then the pressure to the temperature.

The molecules exert a pressure on the walls of the container because of the continual bombardments of the walls. If the molecules have mass m, then one moving with velocity v perpendicular to a wall will have momentum mv. Since the collisions are elastic it will exactly reverse its direction after hitting the wall, and the total change in momentum will thus be $2mv$. The force on the wall is given by the rate of change of momentum (Newton's second law, page 29). Thus we need to determine the number of times per second that a wall is hit by a molecule. This is most easily achieved by considering the gas to be contained in a cubic vessel with sides of length L (Fig. 9.1). Any given molecule will have a velocity C in a random direction, which will have components v_x, v_y and v_z parallel to the sides of the vessel. The time taken to travel between the faces perpendicular to the x-direction will be L/v_x and thus there will be $2L/v_x$ seconds interval between successive hits on one face (across the cube and back again). The number of times per second that one of these faces it hit per second by this one molecule is therefore $v_x/2L$ times/second.

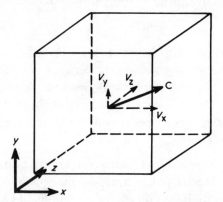

Fig. 9.1 Cube of side length L. A particle with velocity C is illustrated, whose components are v_x, v_y, v_z.

The *rate of change of momentum* is given by the change in momentum in one collision times the number of collisions per second and is:

$$2mv_x v_x/2L \; = \; mv_x^2/L \qquad (9.1)$$

The force on the wall due to this one molecule is given by this rate of change of momentum.

$$\text{Force} = mv_x^2/L \tag{9.2}$$

The total force exerted by the molecule is the sum of the forces on all six walls. Two of these walls are in each direction and therefore the total force is

$$\text{Total force} = 2m(v_x^2 + v_y^2 + v_z^2)/L \tag{9.3}$$

By Pythagoras's theorem $C^2 = v_x^2 + v_y^2 + v_z^2$

$$\text{Total force} = 2mC^2/L \tag{9.4}$$

The pressure is the force per unit area

$$\text{Pressure } (P) = \text{Force}/6L^2 = 2mC^2/6L^3 = mC^2/3L^3 \tag{9.5}$$

L^3 is the volume V of the vessel

$$PV = \tfrac{1}{3}mC^2 \tag{9.6}$$

This equation is as far as Newton's Laws takes us.

9.2   Thermal energy

We now introduce our sixth assumption that the kinetic energy of the molecules $(\tfrac{1}{2}mC^2)$ is proportional to the absolute temperature T.

$$\tfrac{1}{2}mC^2 = \tfrac{3}{2}kT \tag{9.7}$$

The constant of proportionality in this expression (k) is known as Boltzmann's constant. It is the *constant of proportionality between the kinetic energy of a molecule and the temperature*, and is a very important constant. The factor 3/2 is introduced for the not immediately obvious reason that the contribution to the total kinetic energy is $1/2kT$ for each degree of freedom of the molecule. The degrees of freedom that you need to know about are:

1. One for each of the three possible directions of translational movement for a molecule. Thus a molecule restricted to move in only one direction has $\tfrac{1}{2}kT$ of kinetic energy. In the case we have considered the molecule to be free to move throughout the cube and thus have three degrees of freedom and $3kT/2$ of kinetic energy.
2. In diatomic or larger molecules there will in addition to translational motion be the possibilities of the molecule rotating and also of the different atoms of the molecule vibrating. For large molecules of n atoms the *total* number of degrees of freedom is $3n$, and thus the thermal energy is $3nkT/2$.

This energy that a molecule possesses because of the temperature of the medium is known as *thermal energy*. Although we have derived the above relationships in terms of gases, the molecules and atoms in solids also have thermal energy. The average

amount of thermal energy that any particular atom possesses in a solid or liquid
will be of the order of kT. Generally of course the gross properties of a material are
measured, in which case a more useful quantity of energy is the thermal energy of
a gram mole of molecules. In one gram mole there are Avagodro's number N molecules ($N \doteq 6 \times 10^{23}$), and thus the constant of proportionality between energy and
temperature is Nk for one gram mole. This value (Nk) is called the *gas constant R*.

Returning to the problem of the gas we have

$$PV = \tfrac{1}{3}mC^2 \text{ and } \tfrac{1}{2}mC^2 = \tfrac{3}{2}kT$$

giving

$$PV = kT \text{ for a single molecule} \tag{9.8a}$$

$$PV = NkT = RT \text{ for one g-mole} \tag{9.8b}$$

$$PV = nRT \text{ for } n \text{ g-mole.} \tag{9.8c}$$

This is the equation of state of a perfect gas. Many physical phenomena are dependent
upon thermal energy for their effects. For example, rubber elasticity (page 60) and
thermal noise (page 172) are two mentioned in this booklet. Activation energies of
chemical reactions are another. The theoretical treatments of all these subjects make
use of the magnitude of thermal energy, and the resulting equations contain kT or
RT to take account of the magnitude of thermal energy.

9.3   Distribution of energy and velocity

The previous section has discussed the kinetic energy of molecules in terms of the
temperature of the system containing the molecules. However, at any instant of
time not all molecules possess exactly this amount of energy; the thermal energy is
the average energy of the molecules of the system. Some molecules will have energies
greater and some energies less than this value. Is there any expression giving this
distribution?

The distribution of energies is known as the *Boltzmann distribution* determined
from the Boltzmann law which states:

In a system depending for its energy upon thermal energy, the fraction of the molecules that have *energy E or greater* is

$$\text{Fraction} = n_{E+}/N_0 = \exp(-E/kT) \tag{9.9a}$$

in which n_{E+} is the number of molecules whose energy is E or more and N_0 is the
total number of molecules in the system. If the energy is entirely kinetic energy of
translational motion then the *distribution of molecular velocities* is given by

$$N(C) = AC^2 \exp(-\tfrac{1}{2}mC^2/kT) \tag{9.10}$$

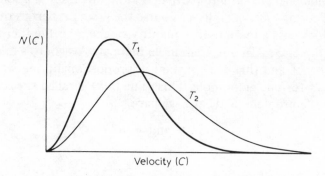

Fig. 9.2 Distribution of velocities according to Boltzmann's Law. $N(C)$ is the number of particles with velocity C. The distribution is shown for two temperatures, T_2 being greater than T_1.

This curve has the form shown in Fig. 9.2. $N(C)$ is the fraction of molecules whose velocity is C and A is the constant of proportionality.

9.4   Boltzmann distribution for a system containing discrete energy levels

Equation 9.9a was given in a form applicable for the situation in which all values of the energy are allowable. However, as is discussed in Chapter 14, at the molecular level only discrete levels of potential energy are allowable; for example, an electron bound to a molecule cannot have any value of energy — its energy at any instant of time must be one of a set of discrete levels $E_0, E_1 \ldots . E_n \ldots .$ Under these circumstances the appropriate form of the distribution of molecules, electrons or whatever is being discussed amongst the discrete energy levels is

$$\frac{n_j}{n_k} = \frac{\exp(-E_j/kT)}{\exp(-E_k/kT)} = \exp(-(E_j - E_k)/kT) \qquad (9.9b)$$

in which E_j and E_k are the energies in the j and k levels respectively.

9.5   Specific heat

We have talked about temperature (in terms of random motion of molecules) but not about heat. An increase in temperature means a greater mean kinetic energy of our material, but how do we cause an increase in temperature. Obviously we must supply energy to the system, and the nature of this energy is known as heat. The relationship between the amount of heat we must supply to unit mass and the temperature increase is known as the specific heat of that material.

specific heat C = (heat supplied/temperature increase) for unit mass. (9.11)

Often we shall be more interested in unit volume than unit mass of a material, in which case we are interested in the *heat capacity* for unit volume. The ratio of mass to volume of a substance is its density ρ

$$\rho = \text{mass/volume}$$

Thus heat capacity of unit volume = density $\times$ specific heat = ρC. Some values of specific heat and density are given in Table 9.1

TABLE 9.1

	Thermal conductivity J/(m s °C)	Density kg/m³	Specific heat (constant pressure) J/(kg °C)
Air	0·024	1·3	10^3
Water	0·6	10^3	$4·2 \times 10^3$
Sea water		$1·03 \times 10^3$	$3·9 \times 10^3$
Ice	0·23	$0·92 \times 10^3$	2×10^3
Fat	0·2		
Fur/feathers	0·035		
Wood	0·15	$(0·5 - 0·8) \times 10^3$	
Steel	50	8×10^3	$0·5 \times 10^3$
Copper	385		

9.6 Heat transfer

How can heat be transferred from one body to another? So far we have only considered heat energy in terms of the random motion of molecules. One obvious way for the transfer of heat is for the kinetic energy of one part of a system to interact directly with that of another part and to transfer kinetic energy by molecular bombardment. Such a mechanism is known as *conduction*, and the laws of conduction concern the mechanical interaction of molecules. Another obvious way is for the bulk movement of a hot material, usually gas (and from a biological point of view nearly always air). This is known as *convection*, and the laws of convection are largely dependent upon the physical differences of hot and cold air such as density.

9.7 Radiation

There is a third way by which heat can be transferred, and to understand this we must consider another form of heat altogether. This is in the form of electromagnetic radiation (visible and infra-red radiation mainly). Conduction and convection, both depending upon the motion of molecules for their heat energy, cannot be used for transferring heat through a vacuum, yet heat very obviously can be transferred. The heat that the earth obtains from the sum is an obvious example of this. This heat is transferred by *radiation* of 'heat waves' which is just electromagnetic radiation, mainly visible and infra-red light (see Chapter 14). The heat you feel when you sit in front of an electric fire, or pur your hand close to a lighted bulb is radiated heat. Its transfer obeys the laws of electromagnetic radiation; thus it can be reflected and refracted. (It is only possible to refract certain wavelengths because there are no known materials capable of refracting the longer wavelengths of infra-red light. Glass, for example is opaque to light of wavelengths longer than about $2\,\mu$m; the energy is then absorbed by the glass.) Examples of reflection of heat are from the curved reflector behind electric fires and from the reflecting surface which is coated onto the glass wall of vacuum flasks in order to reflect the radiant heat transferred through the vacuum.

What determines the amount of heat radiated from a body, or even whether any heat is radiated at all? Obviously the hotter the object the greater the amount of heat radiated. It is found that the rate of radiation of heat from unit area of surface (in terms of energy flow per second) is proportional to the fourth power of the absolute temperature of the object.

$$\text{Radiation} \propto T^4$$

The constant of proportionality is known as *Stefan's constant* σ and the law

$$\text{Radiation/unit surface area} = \sigma T^4 \qquad (9.13)$$

is known as *Stefan's law* or the *Stefan–Boltzmann law*. Stefan's constant has a value of $5{\cdot}69 \times 10^{-8}\,\mathrm{W\,m^{-2}\,K^{-4}}$.

This law is only true for what are termed *black bodies*. The ability of a surface to radiate heat is dependent upon various properties of that surface. The ideal black body is in fact a small hole in the surface of a hollow vessel. All radiation entering this hole is absorbed by the vessel, and this is also the optimal emitter of radiation. Other bodies are less good emitters of radiation, and the radiation emitted from them is

$$\text{Radiation} = e\sigma T^4$$

i.e. radiation from non-black body $= e \times$ radiation from a black body. e is known as the *emissivity* of the body, and the above relationship is known as *Kirchoff's law*. The emissivity must be between 0 and 1. The amount of radiation emitted per unit surface area of a body is known as the *emissive power*, and has units of watt/m^2.

Objects can not only radiate heat, they can also absorb radiant heat, and the law of the absorption is the same as that of emission. The rate of heat absorption is proportional to the fourth power of the absolute temperature of the surroundings, with the same constant of proportionality (Stefan's constant) and the same factor relating the *absorptivity* of black bodies to non-black bodies. This must be the case, since a system reaches equilibrium with all objects in that system at the same temperature. Thus, at equilibrium when the temperature of an object is the same as that of its surroundings the rate of emission of heat equals the rate of absorption. If the object is hotter than its surroundings, that is, at a greater temperature, then the net emission of heat is:

$$e\sigma T_1^4 - e\sigma T_2^4 = e\sigma(T_1^4 - T_2^4) \tag{9.14}$$

where T_1 is the temperature of the body and T_2 that of the surroundings.

9.8   Wavelength of radiated heat

The total amount of heat radiating from a body (the emissive power) is dependent upon temperature. The different wavelengths making up this radiation are also dependent upon the temperature. In a generalized way this is everyday experience. We associate a gentle red with a cooler object than a bright white. A fairly hot object does not seem to have any colour change — this is because its emitted radiation is in the infra-red (outside the visible spectrum). 'Red-hot' implies a higher temperature than this.

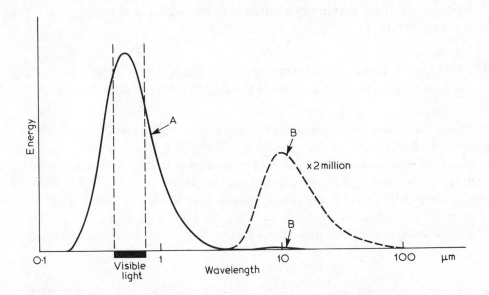

Fig. 9.3 Distribution of wavelength of emission of radiation from a black body. A. The temperature of the sun (6000°K) and B. 300°K (terrestrial temperatures). The solid curves are drawn to the same scale. The dashed cruve is magnified 2×10^6 times.

The form of the distribution of wavelengths emitted from a black body and the way this varies with temperature is illustrated in Fig. 9.3.

The radiation from bodies at two temperatures is shown; the temperatures are chosen to illustrate the distribution of wavelengths in radiation from the sun (*solar radiation*) and from objects at normal terrestrial temperatures (*terrestrial radiation*). Note that terrestrial and solar radiation can be distinguished by their effective restrictions to wavelengths above and below about 2·5 μm respectively. Glass transmits radiation up to about 2·5 μm and is thus an effective means of separating the two forms of radiation. A convenient terminology is to classify wavelengths from visible light up to 2.5 μm as *near infra-red* and wavelengths longer than 2·5 μm as *far infra-red*.

The curve has a maximum emission E_m at a certain wavelength λ_m. There are two laws, known as Wien's laws, which define the dependence of E_m and λ_m upon temperature. These are:

1. The wavelength of the maximum emission varies inversely as the absolute temperature.

$$\lambda_m \, T = \text{constant}$$
$$= 2900 \, \mu\text{m K} \tag{9.15}$$

2. The magnitude of the maximum emitted radiation varies as the fifth power of the absolute temperature.

$$E_m \propto T^5 \tag{9.16}$$

Notice that the total amount of energy emitted is the area under the curves of Fig. 9.3, which, as Stefan's laws says, varies as the fourth power of the absolute temperature.

As stated these laws are empirical in nature. The precise shape of the curves can only be explained in terms of the quantal nature of light. It was, in fact, the inability of classical physics to explain the shape of these curves that lead Planck at the beginning of this century, to postulate the quantal nature of light. This is dealt with briefly in Chapter 15. The conclusion, known as *Planck's law* is that radiation can only be emitted in amounts in which the energy is a multiple of the energy of a basic unit or *quantum*. The energy of one quantum of radiation whose wavelength is λ is

$$\text{quantal energy} = hc/\lambda = h\omega \tag{9.17}$$

in which h is a constant known as Planck's constant (whose value is $6·6 \times 10^{-34}$ J s), c is the velocity of light, λ is the wavelength of the radiation being considered and ω is the frequency of this radiation.

Not only can radiation only be emitted in quanta; it can only be absorbed in quanta. Thus, from the point of view of emission and absorption you must think of radiation being in the form of 'wave packets' of energy $h\omega$ which have to be treated as particles in that they cannot be subdivided.

9.9 Conduction

Conduction is the process whereby heat is transferred from one part of an object to another part of that object by the influence of atoms with a high kinetic energy interacting with adjacent atoms of low kinetic energy by means of molecular bombardment. In solids this takes place without any rearrangement of the atoms. In gases the individual atoms are free to traverse the space of the gas, and thus transference of energy can take place by molecular movement. However the average distance that a gas molecule travels in normal air before colliding with another molecule (the mean free path) is only about $0.1\mu m$. Thus although the molecules can travel through the gas, we can consider conduction of heat in *still* air to occur by the same mechanisms as in solids – by molecular collisions. Nonetheless, in gases the process is sometimes referred to as *static diffusion*.

Experimentally it is found that the amount of heat conducted per second across a piece of material is proportional to the difference of temperature across the material ΔT, the cross-sectional area A, and inversely proportional to the length L.

$$\text{Heat flow per second } \frac{Q}{t} = k\frac{A\,\Delta T}{L} \qquad (9.18a)$$

The constant of proportionality k is known as the *thermal conductivity* of the material. This formula is only true provided that there is no heat loss from the sides of the material (that is, all the heat entering one end is leaving the other). A more useful form of this equation for many purposes is the differential form

$$\frac{dQ}{dt} = kA\frac{dT}{dx} \qquad (9.18b)$$

It is precisely the same as that above but applying to a minute piece of the material. This formula can be used in cases where there is non-uniform heat flow to evaluate the overall transfer of the heat. Some values of thermal conductivity are given in Table 9.1.

9.10 Diffusion

The law pertaining to the diffusion of molecules in the gaseous and liquid phases has exactly the same form as that for heat conduction. The law in this case is known as *Fick's law*.

$$\frac{dn}{dt} = DA\frac{dc}{dx} \qquad (9.19)$$

in which n is the number of molecules diffusing through an area A down a concentration gradient dc/dx. D is known as the diffusion coefficient, with units m^2/s. If

c is in mole/m^3 then n is in mole. Some values of diffusion coefficient of biological interest are given in Table 9.2.

The diffusion coefficient of a gas is proportional to its velocity. At a given temperature the mean kinetic energy of the molecules of any gas is a constant.

$$\text{i.e. } \tfrac{1}{2}m_1 V_1^2 = \tfrac{1}{2}m_2 V_2^2 = \text{const.}$$

Thus

$$\frac{D_1}{D_2} = \frac{V_1}{V_2} = \frac{\sqrt{m_2}}{\sqrt{m_1}}$$

This is known as *Graham's law of diffusion* in gases, and shows quantitatively that light molecules diffuse, through gases, faster than heavy molecules.

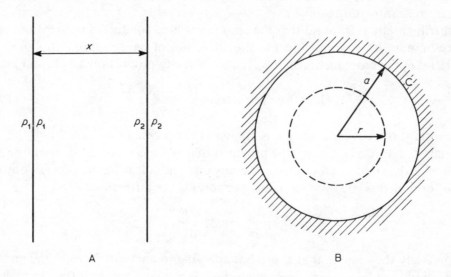

Fig. 9.4A. Diffusion of gas through an aqueous medium. B. Cylindrical tissue of radius a immersed in solution of concentration C.

Example 9.1

A common type of problem in which diffusion is a limiting factor concerns the diffusion of a chemical into a tissue from a source of that chemical (say the blood), under conditions in which the tissue is using that chemical at a constant rate per unit mass. The precise formulation will depend upon the shapes of the tissues. This example is just meant to illustrate the general way of setting about such a problem. Many invertebrates bathe their tissues in haemolymph. Let us consider a cylindrical piece of tissue — this might represent a muscle fibre — as illustrated in Fig. 9.4B. The outer solution maintains a constant concentration C mole/m^3 of the chemical, and the tissue uses this up at a constant rate B mole/m^3/second. Consider the tissue within the radius r (drawn dashed). The volume per unit length of the cell is πr^2, and thus rate of

TABLE 9.2

	% in dry air	Diffusion coefficient D			Solubility in water α			$\alpha D/22\cdot4$
		Air m^2/s	Water m^2/s	Muscle m^2/s	37°C	20°C	0°C	p mole/m s
					m^3 gas/m^3 water/atm			
O_2	20·9	$1\cdot78 \times 10^{-3}$	$1\cdot72 \times 10^{-9}$	$0\cdot75 \times 10^{-9}$	0·024	0·030	0·047	1·85
CO_2	0·03	$1\cdot39 \times 10^{-3}$	$1\cdot46 \times 10^{-9}$	—	0·51	0·85	1·67	33·2
N_2	78·1	$1\cdot90 \times 10^{-3}$	$1\cdot62 \times 10^{-9}$	—	0·012	0·015	0·023	0·87
H_2	5×10^{-4}	$6\cdot34 \times 10^{-3}$	$3\cdot59 \times 10^{-9}$	—	0·016	0·018	0·021	2·57
H_2O (vap)	—	$2\cdot39 \times 10^{-3}$	—	—	—	—	—	—
Glucose	—	—	$0\cdot57 \times 10^{-9}$	$0\cdot28 \times 10^{-9}$	—	—	—	—
Myoglubin	—	—	$0\cdot11 \times 10^{-9}$	—	—	—	—	—
ATP	—	—	$0\cdot3 \times 10^{-9}$	$0\cdot12 \times 10^{-9}$	—	—	—	—
Ca^{++}	—	—	$0\cdot71 \times 10^{-9}$	$0\cdot7 \times 10^{-11}$	—	—	—	—
K^+	—	—	$1\cdot8 \times 10^{-9}$	$1\cdot0 \times 10^{-9}$	—	—	—	—
Na^+	—	—	$1\cdot2 \times 10^{-9}$	$0\cdot6 \times 10^{-9}$	—	—	—	—

usage of chemical within radius $r = B\pi r^2$ mole/second. This must equal the rate at which chemical crosses the boundary of this volume (since the system is in equilibrium). From Fick's Law we see that the rate of transfer of chemical is given by

$$\frac{dn}{dt} = D\,2\pi r\,\frac{dc}{dr} \qquad (9.20)$$

The area of the surface of unit length of this cylinder is $2\pi r$, and the concentration gradient is dc/dr.

Therefore
$$D\,2\pi r\,\frac{dc}{dr} = B\pi r^2$$

$$\frac{dc}{dr} = Br/2D \qquad (9.21)$$

Thus the cencentration gradient is linearly proportional to the radius, and the concentration at a radius r, $c(r)$ is (see Page 13):

$$\int_{C}^{c(r)} dc = \frac{B}{2D} \int_{a}^{r} r\,dr \qquad (9.22)$$

The limits of integration are the external solution parameters and those at our radius r. This gives

$$c(r) = C - \frac{B}{4D}(a^2 - r^2) \qquad (9.23)$$

9.11   Diffusion of gases through liquids

A common biological problem, especially in respiration, concerns the diffusion of a gas originally in air through an aqueous medium. We will illustrate the problem by considering the diffusion of gas across an aqueous layer of thickness x, when the pressure of the gas on one side is maintained at p_1 atmospheres, and that on the other side is maintained at p_2 atmospheres. (See Fig. 9.4A).

The *solubility* of a gas in a fluid (α) is defined as that volume of the gas (as measured at $0°C$ and 1 atmosphere – S.T.P) which dissolves in unit volume of fluid when the gas dissolved is in equilibrium with the gaseous form at 1 atmosphere. If the molecular weight of the gas is M, then the mass of gas dissolved is in equilibrium with the gaseous from at 1 atmosphere. If the molecular weight of the gas is M, then the mass of gas dissolved in $1m^3$ when the external partial pressure of the gas is p is

$$p\alpha M/22\cdot4 \;\; kg/m^3$$

since at S.T.P. M kg of gas occupy $22\cdot4$ m^3. The concentration of the dissolved gas is thus

$$p\alpha/22.4 \text{ mole/m}^3$$

Note that the partial pressure of the dissolved gas is still p atmosphere. The solubilities of various gases are given in Table 9.1.

Thus, if the partial pressures of the gas on either side of the aqueous medium are p_1 and p_2, then the concentrations of the gas at the boundaries of the fluid are $p_1\alpha/22.4$ and $p_2\alpha/22.4$ mole/m^3 respectively. We can therefore immediately apply Fick's law of diffusion to the problem

$$\frac{dn}{dt} = DA\frac{d(\alpha p/22.4)}{dx} = A\frac{\alpha D}{22.4}\frac{dp}{dx} \text{ mole/unit time.}$$

Provided SI units are used throughout, but measuring the partial pressure in atmospheres, then dn/dt is in mole/second.

In our particular example here, in which none of the gas is utilized in the fluid, dp/dx is constant and equal to $(p_1 - p_2)/\Delta x$.

Therefore
$$\frac{dn}{dt} = \frac{A\alpha D(p_1 - p_2)}{22.4 \times \Delta x}$$

We can see that for comparison of the rates of flow of gases through aqueous phases in such situations the product of the solubility product α and the diffusion coefficient D is required. Whereas the ratio of the rates of diffusion of oxygen and carbon dioxide in air is $1.78/1.39 = 1.28$, the ratio of their rates of diffusion in water is $0.024 \times 1.72/0.51 \times 1.46$ at $37°$, equalling 0.055. Carbon dioxide diffuses 18 times more rapidly than oxygen across a fluid, when the source of the gas is in the gaseous state, or when the flow rate is measured with respect to the pressure difference rather than the concentration difference.

Note 1

A variety of different units is used when discussing the diffusion of gases through fluids.

Note 2

The solubility of oxygen and carbon dioxide in blood is considerably greater than that in water due to the presence of blood pigments.

9.12   Mobility

A phenomonon directly related to diffusion is the rate of passage of molecules through fluids due to the action of external forces. One point at which you will meet this idea is in the transport of ions through aqueous solutions due to the action of electrical fields. This is important in electrophysiology which is entirely dependent upon such movement of ions. In this case there is a force F acting on the ions, which then travel through the solution with a velocity v. The reason why v is so much less in liquids than in gases is because of the continual collisions of the ion with the molecules of the liquid. The velocity will be directly proportional to the applied force, and this constant of proportionality is known as the mobility μ, with units $m/N\,s$.

$$v = \mu F \qquad (9.24)$$

This formula is generally true for any molecule moving through a solution under the influence of any force.

There is a precise relationship between the mobility μ of an ion, and the diffusion coefficient of that ion.

$$D = \mu kT \qquad (9.25)$$

Once again we meet thermal energy (kT). The reason in this case is because the greater the kinetic energy of the molecules the greater the capacity of the molecules to diffuse through the solution. The above relationship is sometimes known as *Einstein's relationship*.

Example 9.2

The above relationship between mobility and diffusion coefficient can be used to derive the Nernst equation. Consider the movement of ions across a membrane, within which the relevant diffusion coefficient is D, and mobility is μ. If the electrical potential difference across the membrane is ΔV, and the thickness is Δx, then the force acting on one g-mole (see Chapter 4) is $zF\,\Delta V/\Delta x$. From equation 9.24 above, the velocity of transport across the membrane will be $\mu zF\Delta V/\Delta x$, and the flow of current through unit area at any point will be this velocity times the concentration $C(x)$ at that point.

$$\frac{i}{A} = \mu zFC(x)\frac{\Delta V}{\Delta x}$$

We can also apply Fick's law to the movement of the ions across the membrane.

$$\frac{i}{A} = \frac{1}{A}\frac{dn}{dt} = D\frac{dC(x)}{dx} = \mu RT\frac{dC(x)}{dx}$$

Equating these two equations gives

$$\mu zFC(x)\frac{\Delta V}{\Delta x} = \mu RT\frac{dC(x)}{dx}$$

$$\int\limits_{\substack{\text{across} \\ \text{membrane}}} \frac{\Delta V}{\Delta x}\,dx = \frac{RT}{zF}\int\limits_{C_2}^{C_1}\frac{1}{C(x)}\,dC(x)$$

$$\Delta V_m = (RT/zF)\ln(C_1/C)$$

9.13   Convection

Convection of heat is attained by the bulk transfer of hot material from one place to another. It thus takes place in fluids, both in liquids and gases. The reason for convection is that a hot fluid has different properties to a cool fluid. Its density is normally less, and thus we see that hot fluid will tend to rise with respect to cold fluid. This leads to convection currents, since the replacement of hot fluid by cool as the hot rises will cause movements in many directions and the creation of what we call draughts if the movement is restricted, and wind if not.

9.14   Newton's law of cooling

The study of convection currents is outside the scope of this book. We are interested in the way that animals and plants lose heat, and any laws that we can apply to such heat loss. In fact a very simple law applies, known as *Newton's law.* This states that under constant conditions the rate of loss of heat of a body ($\mathrm{d}Q/\mathrm{d}t = J$) will be proportional to the difference of temperature between the body and the surroundings.

$$J = \frac{\mathrm{d}Q}{\mathrm{d}t} = HA(T_\mathrm{b} - T_\mathrm{s}) \qquad (9.26)$$

H is the constant of proportionality, known as the *heat transfer coefficient*, and A is the surface area of the body.

This is fine. However, what value do we give to H and at what point do we measure T_s? T_s is the temperature of the surrounding sufficiently far from the body that the body is not influencing the temperature to any appreciable extent. H is more difficult to specify, and only applies to any one particular body under any one set of conditions; for example, both wind speed and body shape influence the value of H.

In any real situation the change in temperature at different distances away from a body which is at a different temperature from its surroundings will have the form shown in Fig. 9.5. The temperature profile is illustrated for two different wind speeds.

Under conditions in which there is no wind or convection currents the heat loss will be primarily by conduction (static diffusion). The temperature gradient in the air or other fluid surrounding the object will be low, and there will thus be a thick layer of fluid surrounding the object whose temperature is above that of the surroundings. At higher wind speeds there will be only a thin layer of fluid surrounding the body through which heat loss occurs by conduction since further away the fluid is removed by convection.

We can define a boundary layer as illustrated in Fig. 9.5 in which the slope of the temperature curve neat the surface of the object is extrapolated to the temperature

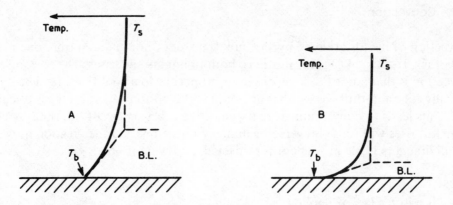

Fig. 9.5 Temperature distribution above an object maintained at a temperature above that of its surroundings in A. low wind, B. high wind. The full curves show the observed temperature profile. The dashed line is the approximation used to obtain the thickness of the boundary layer. It has the same slope as the observed profile near the surface object.

of the surrounding fluid (T_s). The thickness of the boundary layer (B_L) is then the distance required to reach T_s when the temperature gradient has this value. Neglecting radiation for the moment the heat loss by the object can then be found by applying equation 9.18b to the fluid in the boundary layer

$$\frac{dQ}{dt} = kA \frac{dT}{dx} = kA \frac{(T_s - T_b)}{B_L}$$

The heat loss per unit area is thus

$$\frac{J}{A} = \frac{1}{A} \frac{dQ}{dt} = \frac{k}{B_L} (T_b - T_s)$$

which is the same as equation 9.26, with a value for the heat transfer coefficient per unit area

$$H = k/B_L \qquad\qquad (9.27)$$

9.15   Boundary layer thickness

The boundary layer was defined above in terms of the temperature distribution away from the surface of an object. To a good approximation the width of the boundary layer defined in this way will be the same as that obtained by measuring the distribution of fluid velocity above the surface of the object (this was discussed in Chapter 6).

As in the discussions on fluid flow past a surface we must consider whether the flow is laminar or turbulent, and for this the Reynold's number of the object is relevant; laminar flow will be obtained below Reynold's numbers of about 3×10^6. This will generally be the situation when we are considering the boundary layer around such things an animals or leaves, but is not the usual situation when considering the boundary layer over the earth's surface.

For laminar flow of air at velocity V m/s past a flat plate of length L m the heat transfer coefficient is

$$H \doteq 4 \times \sqrt{\frac{V}{L}} \text{ J/(m}^2 \text{ s }°\text{C)} \qquad \qquad \text{flat plate}$$

For laminar flow past a cylinder of diameter D the heat transfer coefficient is

$$H \doteq 9 \times (V/D^2)^{1/3} \text{ J/(m}^2 \text{ s }°\text{C)} \qquad \qquad \text{cylinder}$$

These formulae were derived for forced convection, and are not applicable in still air.

In the case of measurements requiring determination of the heat transfer coefficient for heat transfer from the surface of the earth, it is best to measure the temperature gradient at various heights above the earth's surface, and thus to determine the thickness of the boundary layer.

It must be realized that use of the above formulae will at best give approximate answers only. They are useful however, in estimating the comparative effects that size and wind speed will have on the loss of heat by animals and plants.

9.16   Evaporation

The ideas of the kinetic theory described earlier in this chapter are also applicable to liquids, even though some of the basic assumptions can no longer be applied. In particular the idea of thermal energy is still extremely important, although the form of thermal energy is no longer simply translational kinetic energy, but is concerned largely with vibrational energy between adjacent atoms of the liquid.

The distribution of energy amongst the atoms of the liquid is such that there will be a proportion of molecules which have sufficient energy to overcome the forces of interaction which are responsible for holding the atoms together in the form of the liquid. If these atoms are at the surface, and they are moving in the correct direction they will therefore be able to escape from the liquid. Thus there will be a continual loss from the exposed surface of a liquid of atoms which have above average thermal energy. The effect of this will be to lower the average level of the thermal energy of the liquid – this means that the temperature (which is determined by the average thermal energy) will be lowered. Evaporation leads to a loss of heat from the body of the liquid because of this, and this means of losing heat can be very important.

If the region of space available to the vapour above the liquid is restricted, then the liquid molecules that have entered the gaseous phase will be able to return. If the space available is large, and particularly if there is a current of air to remove the particles of the liquid in the air, then there will be no return. The loss of heat will be increased by factors which cause the removal of the energetic particles that have evaporated into the air. Notice that by means of evaporation a liquid can be maintained at a lower temperature than the surroundings.

The energy transferred by evaporating particles is the product of the latent heat of vaporization L_v of that liquid and the molecules leaving the liquid, n

$$\text{Energy lost} = L_v n$$

The *latent heat of vaporization* is the energy required to overcome the forces holding the molecules of the liquid together, and is measured by determining the amount of energy that must be supplied to convert unit mass of liquid at the boiling point of that liquid to vapour at that temperature. Its value for water is $2 \cdot 256 \times 10^6$ J kg^{-1}.

9.17   Relative humidity

The most common form of evaporation is of water from aqueous solutions. The air will always contain water vapour, and the net rate of evaporation will be dependent upon the amount of water in the air. This amount is measured in terms of vapour pressure or, more generally, in terms of *relative humidity*.

Suppose that a beaker of pure water is placed in a small enclosure. There will be a steady evaporation of water from the beaker into the enclosure. Eventually an equilibrium will be reached when the rate at which water evaporates into the air will be equal to the rate at which it returns due to the total amount of vapour in the air. Under these conditions the water vapour is said to be *saturated*, because any increase in the amount of water vapour in the air causes some water to condense. In other words, no more water can be got into the air in the form of vapour (in fact, under certain conditions it is possible to supersaturate the air, but this is an unstable condition, and will not concern us). Under these conditions we describe the vapour pressure as the *saturated vapour pressure*. This will change with temperature. At low temperatures rather little water can be vaporized. It is a common experience for there to be condensation of water onto cold objects brought into a hot room. Anyone with glasses will have experienced the clouding up that occurs on winter mornings on coming inside from the cold. This is due to the vapour pressure of the air inside being greater than the saturated vapour pressure of air at the temperature of the glasses, and the air immediately surrounding the cold glasses is cooled, thereby causing condensation.

Under normal conditions the air will not be saturated, and the relative humidity is defined as:

$$\text{Relative humidity (\%)} = \frac{\text{partial pressure of vapour}}{\text{saturated vapour pressure at that temperature}}$$

Notice that the relative humidity of a given water content in the air will change with temperature, since the saturated vapour pressure is temperature dependent. This leads to one possible way of measuring the relative humidity. If a shiny metal surface is slowly cooled in the air water will start to condense when the temperature of the surface just reaches the temperature at which the water content of the air becomes the saturated vapour pressure. This is known as the *dewpoint*.

The more usual way of measuring relative humidity makes use of the fact that the net rate of evaporation of water from the surface of water will depend upon the relative humidity of the air, and the temperature drop in the water will depend upon the rate of evaporation. The method therefore is to have two mercury-in-glass thermometers, one of which is dry, the other of which has its bulb surrounded by a cotton wick saturated with water. These are mounted in a frame side by side and the frame is then ventilated by fanning or rotation so that there is maximal evaporation from the water soaking the wick. The wet thermometer cools. The two measurements are the two temperatures. The dry thermometer records the air temperature T, the difference in temperature ΔT between the two thermometers measures the lowering of temperature due to the evaporation from the wet bulb. This is sufficient information to determine the relative humidity which is obtained from tables of T against ΔT. The instrument is known as a *psychrometer*, or wet and dry bulb *hygrometer*.

Another form of hygrometer is the hair hygrometer. This makes use of the fact that the length of a piece of animal hair held under constant tension changes according to its water content, which in turn is dependent upon the relative humidity. The instrument has a piece of hair held under tension and a means of measuring length changes.

9.18 Energy budget of damp surface

Consider the surface of, let us say, wet sand. Under any set of climatic conditions this will be at a constant temperature. What are the contributions tending to make that surface layer gain heat, and what tends to make it lose heat.

Energy influx E_{in}

This will be entirely in the form of radiated heat if we make the assumption that the surface is warmer both than the air above and the ground below the surface.

$$E_{in} = \alpha_s J_s + \alpha_L J_L$$

where α represents the absorbance (or emissivity) of the ground to short-wavelength (s) and long-wavelength (L) radiation. We have included two terms to make the point that the absorbance is different for different wavelengths, and also to indicate that there are two main sources of radiation; the sun which has mainly short-wavelength radiation (see Fig. 9.3) and terrestrial sources which are mainly long wavelength. α will have values in the region of 0·9.

Energy efflux. E_{out}

There will be three different contributions to the heat lost from the surface layer. The first is heat radiated, α obtained from the Stefan–Boltzmann law

$$\text{Radiated energy} = \alpha_L \cdot \sigma \cdot T^4$$

Heat will be conducted away, both upwards into the air and downwards into the ground. These will both be subjected to Newton's law.

$$\text{Conducted energy loss} = H_d (T - T_g) + H_u (T - T_a)$$

in which H_d and H_u are the upward and downward heat transfer coefficients, T_g is the temperature of the ground well below the surface, and T_a is the air temperature.

Heat will also be lost from the surface due to evaporation. The total lost in this way will be

$$\text{Evaporation energy loss} = L_v J_{evap}$$

in which L_v is the latent heat of vaporization of water (this is a measure of the amount of energy required to vaporize unit mass of water) and J_{evap} is the rate at which water leaves the surface.

Thus $\qquad E_{out} = \alpha_L \sigma T^4 + H_d (T - T_g) + H_u (T - T_a) + L_v J_{evap}$

9.19 Questions

1. In one proposed mechanism of muscle contraction the A and I filaments intersect via cross bridges composed of two regions (i) an elastic link and (ii) a contractile link which can take up the two conformations shown:

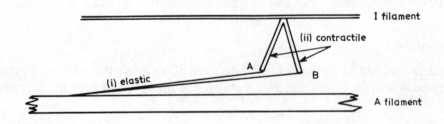

Fig. 9.6

If the stiffness of each elastic link is 4×10^{-4} N/m, the extension caused by the transition between the two conformations of the contractile link is 10 nm, what is the proportion of cross bridges in each of the two states when (a) the elastic link is unstretched with the contractile link in state A, and (b) the elastic link is stretched by 5 nm when the contractile link is in state A?

What is the difference in energy in kJ/mole between states A and B in case (a) above?

cf. Question 5, Chapter 4.

2. A hypothetical animal can be considered to be a cylinder 2 m long and 30 cm in diameter. At night its surface temperature is found to be 25°C when the surrounding temperature is 5°C. What is the net heat loss from the animal due to radiation assuming that it has an emissivity of 0·9. What thickness layer of fur of thermal conductivity 0·035 J/m s °C is required to reduce the heat loss by a factor of 2.

3. A spherical aquatic animal depends upon diffusion from its outer surface for its O_2 supply. If its oxygen consumption is 1 ml/gm wet weight/hour, what is the maximal radius the animal can have if the partial pressure of the O_2 at the centre of the animal can never fall below 1/100 atmosphere. (The logic of example 9·1 is useful here.)

4. A species of animal lives on the bottom of a pond 10 m deep. The individual animals weight 0·01 N and have an oxygen consumption of 0·5 ml O_2/gm wet weight/hour. What is the maximum surviving density of animals possible (in number/m^2)?

5. With each breath your expired air contains about 7% water vapour. What power output does this represent in terms of heat loss due to vaporization of the water?

10 Electricity and Magnetism

This is a 'summary' chapter, stating the basic definitions and laws of electrostatics, current electricity and electromagnetism, without long discussion.

10.1 Charge

The occurrence of charged particles is known from the effect they have on one another. Two types are found, termed *positive* and *negative*, with the properties that like charges repel one another and unlike charges attract one another. The forces of attraction or repulsion obey the law

$$\text{Force} = \frac{1}{4\pi\epsilon} \frac{Q_1 Q_2}{r^2} \tag{10.1}$$

Q_1 and Q_2 are the magnitudes of the two charges, ϵ is a constant known as the absolute permittivity of the medium between the charges and r is their separation. If the sign of the charges is the same then the force is repulsive, if different then attractive. The permittivity of vacuum is known as ϵ_0 and has the value

$$\epsilon_0 = 8\cdot85 \times 10^{-12} \text{ C}^2 \text{ m}^{-2} \text{ N}^{-1}$$

The permittivity of other materials is a factor greater than this, this factor being a property of the material and known as the relative permittivity ϵ_r or dielectric constant. $\epsilon = \epsilon_0 \epsilon_r$

The unit of charge is the *coulomb*, and is defined as that charge which when placed 1 metre from an identical charge repels it with a force of $8\cdot99 \times 10^9$ Newton. A rather more easily grasped concept of what actually constitutes one coulomb can be gleaned from the fact that one g-mole of monovalent ion carries a charge of 96 500 coulombs. This constant is known as the *Faraday*. It follows that one coulomb is roughly speaking that charge carried by $10\,\mu$mole of monovalent ion (or $5\,\mu$mole of divalent ion etc.).

The main carriers of charge from a biologist's requirement are 1. the electron which has a single negative charge and is the main carrier of charge in metals and 2. ions, which are charged atoms or molecules, and which are the carriers of charge in solution (e.g. in cells). Ions can carry more than one charge. The number they carry is called *valency* of that ion, and is given the symbol z.

10.2 Charge induction

If an uncharged conductor is brought into the vicinity of a charged conductor, then charges of the opposite sign to those of the charged conductor will be attracted towards it, and those of opposite sign repelled away. This will lead to a charge distribution on the uncharged conductor. If a conducting lead is attached to the uncharged conductor then the repelled charges will pass along it, and the initially uncharged conductor will appear charged. The amount of charge will be equal in magnitude to that of the charged conductor but of opposite sign.

Thus there will be interaction between conductors. This is of importance to the neurophysiologist especially, since electrodes which are supposed to be recording from a preparation will be able to pick up unwanted signals from neighbouring apparatus. This effect is known as *electrostatic interference*.

10.3 Potential

The force between charges means that work has to be done to move one charge relative to another, and a system of charges is therefore a source of potential energy (P.E.). The P.E. of two charges separated by a distance d was evaluated in Chapter 4 and found to be

$$\text{P.E.} = \frac{1}{4\pi\epsilon}\frac{Q_1 Q_2}{d} \tag{10.2}$$

This was determined by finding the work done in moving a charge from an infinite distance away to this separation. An infinite distance away is a theoretical concept, and a more useful parameter is the difference of potential energy between two points. This is said to be 1 volt when the amount of work done in moving one coulomb between those points is one joule. This, the unit of potential is

$$1 \text{ volt} = 1 \text{ joule/coulomb} \tag{10.3}$$

Although this is the fundamental definition of voltage, it is much easier to think of a potential difference as the driving force which tends to cause charges to move. Negative charges are attracted towards points at positive potential, and positive charges to negative potentials.

10.4   Electric field

We have been talking about the potential of a conductor, and potential differences
as the difference in potential between two conductors. Let us think about this
second case a little more. If we have two conductors separated by a region of non-
conductor (such as air) then the potential will change as we go from one conductor
to the other. Any point in the region between the two conductors will have its own
value of the potential. The precise way in which the potential changes with distance
needs to be evaluated in every case – for the parallel plates illustrated in Fig. 10.1
the change is linear as shown in the diagram. For an isolated charge the curve was
illustrated in Fig. 4.1 in Chapter 4. In that chapter we saw that the force on the
charged particle was the 'potential gradient' at that point. In electrostatistics the
force on a charged particle at a point is known as the *electric field*. Since force is a
vector quantity it follows that the electric field is also

$$\text{Electric field} \ = \ \frac{\mathrm{d}V}{\mathrm{d}x} \ \text{volts/metre} \tag{10.4}$$

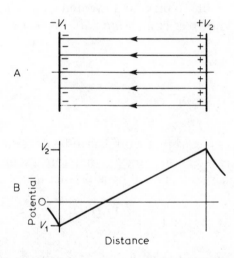

Fig. 10.1 A. lines of force between two charged plates. B. Corresponding potential between plates.

It is customary to draw vectors representing the direction of the electric field, the
number of vectors per unit cross section representing the magnitude of the field.
These are known as *lines of force*. From an isolated charge they are as shown in Fig.
10.2. A wire carrying a current will be charged and the lines of force from such a
wire also radiate out.

   A good conductor is one with a low resistance. The potential of a good conductor
is the same at all points on it, and thus there is no potential gradient. In other words
there is no electric field across a conductor. This fact is of extreme importance for

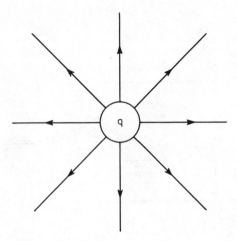

Fig. 10.2 Lines of force from an isolated charge q.

the neurophysiologist since it follows that charges outside a metal box can have no influence inside the box. Other ways of saying this are to say that lines of force end on a good conductor, or that the field inside a box due to influence outside is zero. One common difficulty experienced in recording the extremely small electrical effects in nerves and other excitable cells is that of interference from external electrical apparatus. One way of reducing these effects is by screening the preparation inside an earthed metal box, made from copper or some other good conductor (see Fig. 10.3).

10.5   Current

The movement of charges due to a potential difference constitutes an electric current. One unit of current (the *ampere*) is defined as a rate of flow of one coulomb per second

$$1 \text{ ampere} = 1 \text{ coulomb/second} \tag{10.5}$$

$$I = \frac{dQ}{dt} \tag{10.6}$$

If you have difficulty in understanding the flow of electricity and the various terms, then use the analogy of water flow. The driving force in this case is the pressure of the water, equivalent to the driving force of electric charge, the voltage. The flow of water is equivalent to the current, and the water itself equivalent to the charge.

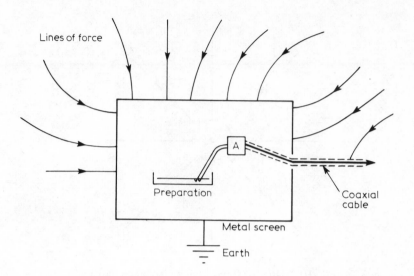

Fig. 10.3 Effect of screening a preparation inside a metal box. External lines of force end at the box.

10.6   Conductors, insulators and semiconductors

In terms of their ability to conduct electricity materials group themselves into the three classes: conductors, non-conductors (or insulators) and semiconductors. In solids, as opposed to liquids, electricity is carried by electrons. The electrons around an isolated atom or molecule are normally fairly tightly bound to that atom or molecule. In aqueous solution certain materials ionize; that is, the material breaks down into oppositely charged particles which are free to diffuse throughout the solution. The negatively charged particles carry one or more extra electrons, the positively charged ones carry one or more too few electrons and hence have a positive charge. In solids the interaction between the atoms making up the solid is such that the possibility exists of the valency electrons being sufficiently free that they can diffuse freely around the solid, in some ways analogously to the ions in solution. The classification of solids depends upon the ease with which these electons can become 'free'. The electrons that do free themselves from their particular atoms exist in what is known as the *conduction band.* This is an energy level within the solid. The normal energy level of the electron on an isolated atom is what is known as the *ground state* of the electron. This is illustrated in Fig. 10.4. Given sufficient energy the electron can leave that state and jump to higher energy levels. The ability to change energy levels is provided by the thermal energy ($3/2 \times kT$) of the electrons. If the energy difference between the ground state of the electron and the conduction band is much less than thermal energy (about 4 kJ/mole) then the conduction band will be readily accessible to the electrons and we have a conductor. If the energy difference is much greater than thermal energy there will be no way for the electrons to reach

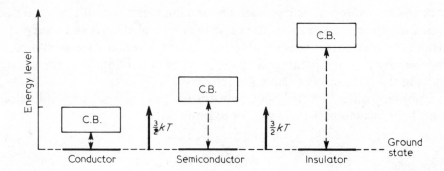

Fig. 10.4 Energy difference between ground state of valency electrons and the conduction band (C.B.) in conductors, semiconductors and insulators. The value of thermal energy $(\frac{3}{2} kT)$ is also shown.

the conduction band and we have an insulator. However, if the energy difference is about equal to thermal energy, then we have the interesting condition that at any instant of time a certain fraction of the electrons will be in the conduction band and the remainder in the ground state. The fraction of electrons in the conduction band at any instant of time can be obtained from the Boltzmann distribution and is:

$$N_c = N_0 \exp(-E/kT) \tag{10.7}$$

in which N_c is the number of electrons in the conduction band, N_0 is the total number of valency electrons and E is the energy difference between the ground state and the conduction band.

It so happens that semiconductors tend to be Group 4 materials (silicon and germanium are most commonly used). The conduction electrons can be increased by the addition of 'impurities' into this material. Group 5 materials added in small concentration provide excess electrons, whereas Group 3 materials provide extra space for electrons from the conduction band and give rise to 'holes', which are equivalent to positively charged 'electrons' in the conduction band. By this means two different types of semiconductor material can be made, called p and n for Positive and Negative.

The usefulesss of semiconductors from the point of view of making transducers (see Chapter 11) is that many effects can cause the number of electrons in the conduction band to change — the effects in principle contribute a certain amount of extra energy and this causes a change in the number of conduction electrons. This change in electron number in the conduction band is apparent as a change in resistance of the material, and many semiconductor transducers are variable resistance devices in which the resistance is dependent upon some external parameter. The most obvious of these is in the temperature sensitive *thermistors.* It follows directly from the equation above that if the temperature is increased, the thermal energy of the electrons will be increased and the number of conduction electrons increases, thereby reducing the resistance of the material. This is what happens in a thermistor. The effect is also responsible for the fact that many semiconductor devices have very high temperature drifts − a strong detraction from the use of transistors in the early days of semiconductors.

However many other factors beside temperature can increase the conduction electrons: light, magnetic field, strain for example.

10.7   Impedance

A potential difference between two points represents a tendency for current to flow between those points. Whether or not any current actually does flow depends upon the impedance to current flow between those points. In general terms we define

$$\text{Impedance} \ = \ \frac{\text{amplitude of voltage}}{\text{amplidute of current}} \tag{10.8}$$

The simplest form of impedance is resistance.

10.8   Resistance and conductance

A resistance is a material which obeys *Ohm's law* which states that the current through that material is directly proportional to the voltage difference across that material

$$V \propto I$$

The constant of proportionality is the resistance R

$$V \ = \ IR \tag{10.9}$$

The unit of resistance is the *Ohm* and is that resistance which allows a current of 1 ampere to pass when the voltage difference across it is 1 volt.

$$1 \text{ ohm} \ = \ 1 \text{ volt/ampere.}$$

It is often useful to talk about the inverse of resistance which is called *conductance* (G).

$$G \ = \ 1/R \tag{10.10}$$

$$I \ = \ VG \tag{10.11}$$

An increase in conductance is equivalent to a decrease in resistance. You will find that neurophsiologists generally talk about the conductance of membranes rather than their resistance, mainly because they think that the ease with which charges cross membranes is dependent upon the number of pores or the number of carrier molecules in the membrane. In this case the conductance will be directly proportional to this number.

The symbol used to represent a resistance is:

Fixed   Variable

Energy is lost when a current flows between two points at different potential. Since the potential difference is a measure of the amount of work done in transferring unit charge it follows that the work done in transferring a charge Q through a potential difference V is QV.

The rate at which work is done is known as *power*.

$$\text{Power} = \frac{d(\text{work})}{dt} = \frac{d(QV)}{dt} = V\frac{dQ}{dt} = VI$$

This represents the rate of loss of energy in a resistor. From Ohm's law we see that

$$\text{Power in a resistor } (R) = VI = \frac{V^2}{R} = I^2R \tag{10.12}$$

The unit of power is a *watt*.

Notice that this power shows itself as a heating effect in the resistor, and thus the energy is lost from the system.

10.9   Capacitance

This is a second form of impedance, and one of great importance in the consideration of membrane properties.

Any two conducting materials, separated by a gap containing a non-conducting material, forms a *capacitor* (often also called a *condenser*). To the physicist the conducting materials are generally metals. To the biologist they are often the two solutions on either side of a membrane; the membrane forms the insulation separating the solutions. The property of a capacitance of interest is its ability to store charge.

If a potential difference is applied across the two plates of a capacitor, then positive charge accumulates on one of the plates, and an equal amount of negative charge on the other. The amount of charge (Q) stored will depend upon the applied voltage. It is found that the amount of charge is directly proportional to the voltage.

$$Q \propto V$$

The constant of proportionality is called the *capacitance C*

$$Q = CV \qquad\qquad\qquad (10.13a)$$

$$C = Q/V \qquad\qquad\qquad (10.13b)$$

The unit of capacitance is called the farad, and is defined by

$$1 \text{ farad } = 1 \text{ coulomb/volt} \qquad\qquad\qquad (10.14)$$

The magnitude of the capacitance (the capacity to store charge) depends upon the physical shape and dimensions of the capacitor.

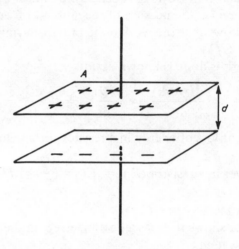

Fig. 10.5 Parallel plate capacitor. Area of plates, A; separation d.

Notice that the larger the value of the capacitance the larger is the amount of charge stored for a given voltage difference between the plates. We can intuitively obtain some idea of the parameters which determine the magnitude of the capacitance. Let us consider the case of a parallel plate capacitor. The potential difference between the plates will be affected by the density of charge on the plates; the denser the charge, the greater the potential difference. Thus we can see that the larger the area of the plates the larger will be the value of the capacitance. For a given potential difference, the larger the area of the plates the more charge will be stored. On the other hand the further the plates are apart the smaller the amount of charge that is stored for a given potential difference between the plates. The reason for this is that the effect of a store of negative charge close to a store of positive charge is to neutralize the potentials due to those charges to some extent. The force on a charge some way away from the plates will be the sum of that due to the positive charge and that due to the negative. These will tend to cancel one another (being in opposite directions), except very close to the plates. This neutralization becomes smaller the further the separation of the plates. Thus the larger the separation between the plates, the smaller is the capacitance. For the parallel-plate capacitor of Fig. 10.5 the capacitance is defined as

$$C = \frac{\epsilon A}{d} \qquad (10.15)$$

in which the area of the plates is A, the separation is d and ϵ is the permittivity of the material between the plates. If the area is measured in m^2 and the separation in m, then C is obtained in farads.

If a steady potential difference is maintained between the plates of a capacitor, then the amount of charge stored will be constant. However, if the potential changes, then the amount of stored charge changes. It does this by leaving or reaching the plates via the electrical connections to those plates. These electrical connections carry a flow of charge under these conditions of changing potential i.e. when the potential is changing there is a current flow onto one of the plates and leaving the other. Although no actual charge crosses the insulating gap between the plates, a current measuring device placed in the connecting leads would 'see' current in both leads and would not realise that there was no charge crossing the gap.

We can easily determine the relationship between current and potential for a capacitor. Our capacitance was defined by the relationship

$$V = \frac{Q}{C} \qquad (10.13c)$$

If we differentiate this expression we get:

$$\frac{dV}{dt} = \frac{d(Q/C)}{dt} = \frac{1}{C}\frac{dQ}{dt} = \frac{1}{C}I$$

$$I = C\frac{dV}{dt} \qquad (10.16)$$

We have used equation 10.6 to equate dQ/dt to I. This is the relationship equivalent to Ohm's law for a capacitor. We see, as we expected, that the current is proportional to the rate of change of potential. Thus if the rate of change of potential is very rapid the current will be very large. It will appear under these conditions that the impedance to current flow is very small. If the rate of change of potential is very slow however, the current will be very low, and it will appear that the impedance is very large. The capacitor is acting as an impedance whose value depends upon the rate of change of potential.

A convenient way to express this is in terms of the impedance to a sinusoidal input voltage across the capacitor. If the input voltage has a waveform $V = V_0 \sin(\omega t)$ (the meaning of ω is discussed in Chapter 12) then the current

$$I = C\frac{dV}{dt} = C\frac{d(V_0 \sin(\omega t))}{dt} = C\omega V_0 \cos(\omega t)$$

i.e. $$I = \omega C V_0 \cos(\omega t)$$

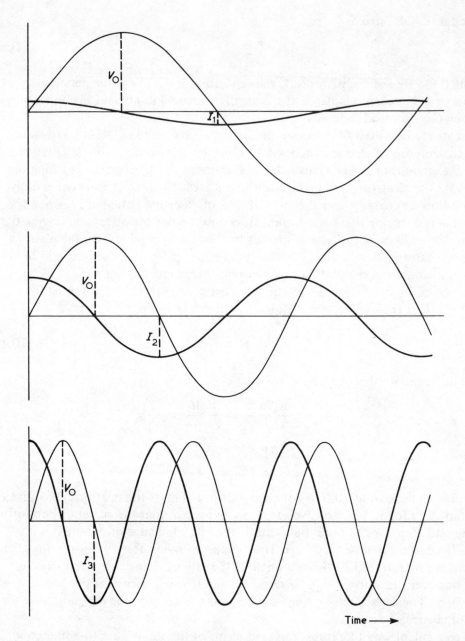

Fig. 10.6 Current across a capitance at three different frequencies, applying a constant amplitude voltage. At high frequencies the current amplitude is much greater than at low frequencies i.e. $I_3 > I_2 > I_1$. The impedance to current flow is thus frequency dependent.

The current waveform for a constant amplitude voltage waveform at three different frequencies is shown in Fig. 10.6. Notice first that the peaks of the current waveform occur when the voltage is zero. The current is said to be $90°$ out of phase with the voltage. This we can tell from the mathematical equations, since the voltage is a sine wave and the current a cosine wave. The *amplitude* of the waveform is height of one of the peaks. The impedance of a capacitance is defined for sinusoidal input waveforms, by analogy with Ohm's law as the ratio.

$$\text{Impedance} = \frac{\text{amplitude of voltage}}{\text{amplitude of current}}$$

$$= \frac{V_0}{\omega C V_0} = \frac{1}{\omega C} \qquad (10.17)$$

Thus we see that the impedance of a capacitor is inversely proportional to the magnitude of the capacitance and the frequency of the imposed waveform.

The store of charge on a capacitor is an energy store. The amount of energy stored can be evaluated by determining the amount of work that had to be done to charge up the capacitor. Let the final charge and potential of the capacitor be Q and V. At some instant during the charging-up process the charge on the capacitor was a fraction of α of the final charge, and the potential was therefore αV. The work done in bringing up the next fraction of charge $(d\alpha Q)$ is $V(d\alpha Q) = VQ\alpha d\alpha$. The total amount of work is found by integrating this expression from $\alpha = 0$ to $\alpha = 1$

$$\text{Energy} = \int_0^1 VQ\alpha d\alpha = QV \int_2^1 \alpha d\alpha = \tfrac{1}{2}QV$$

Using the relationship $C = Q/V$ we obtain the following expressions:

$$\text{Energy stored in a capacitor} = \tfrac{1}{2}QV = \tfrac{1}{2}CV^2 = \tfrac{1}{2}Q^2/C \qquad (10.18)$$

10.10 Magnetic effects

We know about the existence of charged particles because of the effect they have on one another. In an exactly analogous way we know about the existence of magnetic poles because of a force of interaction between them. The law relating the magnitude of the force to the magnitude of the poles (m_1 and m_2) and their separation (r) is also an inverse square law

$$F \propto \frac{m_1 m_2}{r^2} \qquad (10.19)$$

Thus in the vicinity of one magnetic pole we see that there is an interaction on other poles, and we define the *magnetic field* at a point in space, in the same way as we defined an electrostatic field, as the force on a unit pole at that point.

One difficulty with magnetic poles that we did not experience with electrostatic charge is that we cannot obtain an isolated pole. Any piece of material with one pole will also carry one of opposite polarity. This is a magnet. The two polarities are termed

north and south rather than plus and minus and they are distinguished according to
whether the force is one of attraction or one of repulsion.

10.11   Magnetic field due to a moving charge

It is found that a wire carrying an electric current creates a magnetic field. For a
short length dx of wire carrying a current I the magnetic field a distance r away as
shown in Fig. 10.7 is

$$dH = \frac{I \sin\theta}{4\pi r^2} dx \tag{10.20}$$

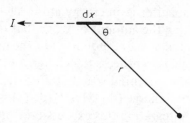

Fig. 10.7 The direction of the field is perpendicular to both the wire and the vector **r**. Thus in the
figure the field is coming out of the paper.

10.12   Force on a moving charge in a magnetic field

It is also found that a charge moving in a magnetic field experiences a force. The
direction of the force is perpendicular to both the direction of movement and the
direction of the magnetic field, and has magnitude

$$F = \mu I H \sin\theta \, dx \tag{10.21}$$

in which μ is the permeability of the medium (very large for magnetic materials) and
θ is the angle between the magnetic field and the direction of current flow. (μH is
known as the *magnetic induction* (denoted by B), and since the force is dependent
on B it is normal to draw lines of force as lines of magnetic induction. These tend
to congregate in material of high μ, shielding from magnetic effects can be obtained
by surrounding the preparation by a box of material with high μ.) This formula
related the force to a current flowing in a wire. For a charge moving with velocity
v the expression is

$$F = \mu Q H \sin\theta \, v \tag{10.22}$$

where θ is now the angle between the field and the direction of movement.

Note. Those interested in vector notation will recognize that the three equations above could all be written more neatly as vector products (see page 27).

10.13   Interaction between two wires each carrying a current

From the above two sections we can see that two parallel wires, each carrying a current will exert a force on one another. The current in one wire will create a magnetic field as shown in Fig. 10.8. This field is perpendicular to the flow of current in the other wire, which will therefore be acted on by a force. This force will be in the direction of the other wire — attractive if the currents are flowing in the same direction, and repulsive if in opposite directions. This effect is used to define the magnitude of the *ampere*, the unit of current. One ampere is that current which exerts a force of 2×10^{-7} newtons/metre on an identical current, when the two currents are flowing in long parallel wires one metre apart in vacuum.

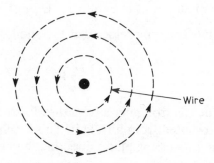

Fig. 10.8 Magnetic field (dashed lines) around a wire carrying a current.

10.14   Creation of magnetic fields

Although permanent magnets produce magnetic fields, an often more useful way of their production is with a solenoid. This is simply a coil of wire.

Application of equation 10.20 to a long coil of wire shows that the field in the centre of a solenoid is

$$H = mI \tag{10.23}$$

in which m is the number of turns per unit length in the coil. If the coil is filled with material of high permeability then the field strength will be increased by a factor μ.

10.15   Control of movement of charged particles in apparatus of biological interest

Many instruments make use of the fact that the paths of charged particles can be

influenced by electrostatic and magnetic fields. These include the electron microscope, the cathode ray oscilloscope and the mass spectrometer.

Both the cathode ray oscilloscope (see Chapter 11) and the electron microscope
require a beam of electrons for their operation. These are produced from a heated
filament, which, in a vacuum, emits electrons when sufficiently heated for the
electrons to have sufficient thermal energy to escape. In the absence of any other
effect these electrons would remain in a cloud around the filament. Both instruments
require a beam of electrons. This is produced by an *anode*, a metal plate whose
potential is kept at a high positive level relative to the *cathode*, the source of electrons.
The electrons are thus attracted towards the metal plate by a force, and from Newton's
second law they accelerate due to the action of this force. How fast are the electrons
moving when they reach the anode? Suppose that their velocity when they leave
the electron cloud at the anode is zero. By the time that they reach the anode they
have gained qV joules of energy, where V is the difference of potential between the
two plates and q is the charge on an electron. They will possess this energy as
kinetic energy

$$\text{Energy of electrons at anode} = 1/2mv^2 = qV$$

where m is the electron mass and v its velocity at the anode.

Thus the velocity

$$v = \sqrt{\left(\frac{2qV}{m}\right)} \tag{10.24}$$

In both the CRO and the electron microscope the electrons must be accelerated
towards the anode, but prevented from hitting it. This is achieved with an anode
with a hole in its centre and focussing plates directing the beam through the hole.

Electrons can be considered as electromagnetic waves. Their wavelength λ is related
to their momentum (mv) by

$$\lambda = \frac{h}{mv} \tag{10.25}$$

where h is Planck's constant.

Combining equations 10.24 and 10.25 gives

$$\lambda = \sqrt{\left(\frac{h^2}{2mqV}\right)} = 1\cdot23 \times 1/\sqrt{V} \text{ nanometres (nm)} \tag{10.26}$$

Thus we can control the wavelength of our electron beam by changing the anode
voltage. Electron microscopes usually work with anode potentials of 60–100 kV,
although for some purposes much larger potentials are used. (See Question 1).

The cathode ray oscilloscope requires a method for deflecting the electron beam
in a controlled manner. This can be achieved by either electrostatic or electromagnetic means. Since a charged particle is deflected in the direction of an electric field,
electrostatic deflection is caused by producing an electric field in the direction of

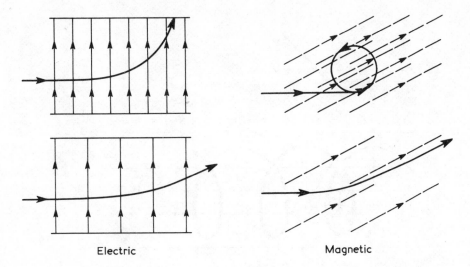

Electric   Magnetic

Fig. 10.9 Path of an electron in strong (upper line) and weak (lower line) electric and magnetic fields.

required deflection by two metal plates whose potential is controlled by the input. Electromagnetic deflection of an electron beam moving with velocity v is caused by a magnetic field which is perpendicular to the direction of deflection. Since the force on the electrons is also perpendicular to their movement (and is still perpendicular to their new direction of movement after they have been deflected) it follows that in a sufficiently large magnetic field (either large in magnitude or volume) the beam will be sent into a circular path. The deflections of an electron beam in large and small electric and magnetic fields is illustrated in Fig. 10.9.

Nearly all modern oscilloscopes use electrostatic focusing. It is easier to obtain uniform electric fields over large volumes than magnetic ones. The best magnetic fields are produced by two tightly packed coils, one either side of the potision at which the field is required, and whose radius equals their separation. This arrangement is known as Helmholtz coils.

The electron microscope depends for its operation upon the ability of fields to focus electrons. This is normally achieved by magnetic fields. We have seen how an electron beam moving at right angles to a magnetic field is caused to take up a circular orbit. Suppose that this electron beam has also got a component of motion in the direction of the field. (Notice that we can represent the velocities parallel ($V_{\parallel}$) and perpendicular ($V_{\perp}$) to the field as vectors — the overall motion is the vector sum of these. In other words we have resolved the overall movement into a component perpendicular to the field which *is* acted on by a force due to the field, and a component parallel to the field which experiences *no* force.) The resulting motion in the magnetic field will be a spiral. You can visualize this by imagining that as the beam in Fig. 10.9 rotates round the circle it is also moving into the paper. The effect is shown in Fig. 10.10.

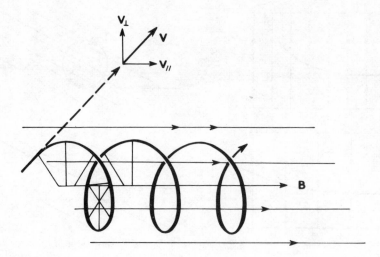

Fig. 10.10 Motion of a charged particle in a magnetic field B. The velocity is initially V with vectors $V_\parallel$ and $V_\perp$.

Now consider the effect of a magnetic field which is parallel to the main direction of the electron beam from the electron gun in an electron microscope. If the beam remains parallel to the field there will be no effect. Suppose that the electron beam is scattered by something put in its way (such as a biological section stained with heavy metal). The scattered electrons will no longer be parallel to the magnetic field. In the absence of the field they would diverge more and more. However, in the presence of the field the scattered electrons all take up spiral motions which not only stops them diverging, it also brings them back to the same point after each complete revolution. In other words the scattered beam has been refocused, and the field has acted as a lens. This is an extremely simple view of the action of magnetic lens, and is meant to illustrate the principles. It is too simple because the image formed in this case is the same size as the object; there has been no magnification. The object and the image were both considered to be in the uniform magnetic field. Focusing can still be achieved when the object is outside the field. In this case the image will also be outside and the action of the magnetic field will be much more analogous to the optical lens. It still achieves the focusing action by producing spiral movement, but the position of focus will not necessarily be after a complete turn of the spiral. You will observe if you use an electron microscope that the image rotates as the magnification is changed. This is because for different magnifications a different amount of spiralling is necessary to achieve focus.

In the light microscope different magnifications were obtained by using lens of different powers. In the electron microscope the power of the electromagnetic lens is dependent upon the magnitude of the magnetic field. Using electromagnets this can be controlled by changing the current flowing through the coils of the

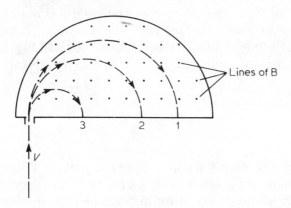

Fig. 10.11 Mass spectrograph. Ionized particles all with initial velocity v enter a chamber with magnetic field B across it. Dashed lines show motion of particles. 1 has smaller value of Q/m than 2, and 2 than 3.

electromagnet. The electromagnetic lens is essentially a solenoid, which produces a field parallel to the axis of the solenoid, but with specially designed pole pieces made from material with a high permeability (soft iron) to create intense fields in the correct position.

One drawback of the electromagnetic lens is the fact that focusing is only good for electrons which have been scattered through *very* small angles. The defocusing effect of the lens for electrons with large angles of scattering is known as spherical abberation (see page 210). In order to stop these electrons with large scattered angles the numerical aperture of the electron lens is made very small, reducing the resolving power considerably (see page 219).

The *mass spectrometer* is an instrument used for identifying the ratio of the charge to mass of charged particles. Since large molecules can be split into ionized components by coating a filament with that substance the spectrometer can be used to obtain information about the constituent components of larger molecules.

The principle of the machine is to accelerate the ions in an electric field, select those with a given velocity v and then pass the resulting beam through a magnetic field which is perpendicular to the direction of motion of the ionised beam. The ions will experience a force BQv perpendicular to their motion and hence go into a circular motion. The centripetal force of the circular motion will be equal to the force due to the action of the magnetic field.

i.e.
$$\frac{mv^2}{r} = BQv$$

$$\frac{Q}{m} = \frac{v}{Br}$$ (10.27

Since the velocity and field strength are known the ratio Q/m can be determined by measuring the radius of the circular motion. This is done by making the particles hit a screen after travelling through a semicircle as shown in Fig. 10.11. Most of the ionized particles will carry a single charge and thus their mass can be determined.

10.16   Question

1. Electron microscopes often work with an anode voltage of 80 000 volts (80 kV). What is the wavelength of the electrons being used? If the numerical aperture is 0·005, what is the resolving power of the microscope (see equation 13.21).

11 Electronic Apparatus

The purpose of this chapter is to provide you with some background to the use of electronic apparatus, mainly from the point of view of the neurophysiologist, but also more generally.

The general type of physiological experiment is one in which some parameter of the working of an animal, a particular organ from an animal or a single cell, is measured as a function of some kind of input. Nowadays the apparatus to make the recording of this parameter will almost always be electrical. The general arrangement for the recording apparatus will be, in block diagram form:

Fig. 11.1 General method of physiological recording.

The transducer is the piece of apparatus which converts (transduces) the energy being measured into an electrical signal. Examples of transducers are: tension transducers for measuring muscle tension (these give a voltage output proportional to the force applied); thermometers for measuring temperature (for ease of recording use is made of those whose output is electrical, such as the thermocouple or thermistor); electrodes for measuring partial pressure of oxygen in fluids. We shall come back to some aspects of transducers later in the chapter.

Some kind of apparatus will be necessary for recording the parameter in a form which the investigator can study. This will be an instrument which accepts an electrical signal, and displays this in some form, either graphically or in the form of a display of numbers (this is called a *digital* display) — the graphical form is sometimes called an *analogue* display. The most important recording apparatus is the cathode ray oscilloscope, but penrecorders are also very useful, and the digital computer is also often used as a device to accept data directly from a transducer.

More often than not the form of the electrical signal obtained from the transducer will not be directly acceptable to the recording apparatus either because its amplitude is too small, or for other reasons we shall meet shortly. In this case an intermediate piece of apparatus must be inserted which corrects this maladjustment, and this is usually an amplifier of one sort or another.

We shall deal with all three parts of this recording assembly: the amplifier, the recording apparatus and the transducer. First, however we must investigate a little circuit theory, making use of the concepts of the previous chapter.

11.1 Voltage drop across two impedances in series

In the last chapter we stated Ohm's law to define a resistance. There are two ways of thinking about Ohm's law. The first is in order to determine the amount of current flowing through a given resistance when a particular potential difference is applied across the ends of that resistance. In this case the current

$$I = V/Z$$

The other way of using Ohm's law is to determine the potential difference between the ends of a resistance if there is a current I flowing through the resistance, in which case

$$V = IZ$$

In Fig. 11.2 a source of potential V_{in} is connected across two impedances in series. A current will flow round the circuit, and the same current must flow through both impedances. We can determine the magnitude of this current from Ohm's law

$$I = \frac{V_{in}}{Z_1 + Z_2}$$

Since this current flows through both impedances we can determine the potential difference between the ends of each impedance, since this p.d. will be (using Ohm's law once again) the product of the current and the impedance.

We shall write V_1 as the potential difference across Z_1 and V_2 that across Z_2.

Then
$$V_1 = IZ_1 = \frac{V_{in}}{Z_1 + Z_2} Z_1 = \frac{Z_1}{Z_1 + Z_2} V_{in} \tag{11.1a}$$

Likewise
$$V_2 = \frac{Z_2}{Z_1 + Z_2} V_{in} \tag{11.1b}$$

Notice that $V_1 + V_2$ equals the input voltage V_{in}. In other words, part of our input voltage is 'dropped' across Z_1 and part across Z_2. Let us consider V_2 for a moment. If Z_2 is very much larger than Z_1 then the voltage difference across Z_2 is very nearly equal the input voltage, whereas if Z_2 is very much less than Z_1 then the voltage across Z_2 will be only a small fraction of the input voltage.

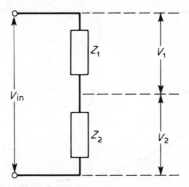

Fig. 11.2 V_{in} connected across two impedances Z_1 and Z_2 in series. V_1 and V_2 are the voltages measured across the impedances.

11.2 Filters

These above relationships are true whether V_{in} is constant or changing, and they are also true whether Z_1 and Z_2 are constant or changing. We saw in the previous chapter that a capacitor is an impedance whose value is dependent upon the rate of change of potential, or for a sinusoidal input, whose impedance is frequency dependent. Suppose, in the above circuit that we make Z_2 a capacitor (condenser) and Z_1 a resistance and that we apply a sinusoidal input voltage as V_{in}. At low frequencies the impedance of the capacitor (equal to $1/\omega C$) is large, and the potential difference between the plates of the capacitor will be nearly equal to the input voltage. At high frequencies the impedance of the capacitor will be low, and the voltage across the capacitor will be a small fraction of V_{in}. If we use the voltage across the capacitor as our output voltage from this circuit we have a device which selectively allows the low frequencies to be outputed, but rejects the high frequencies. This is known as a *low-pass filter*.

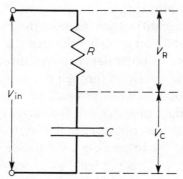

Fig. 11.3 Resistance R and capacitance C in series.

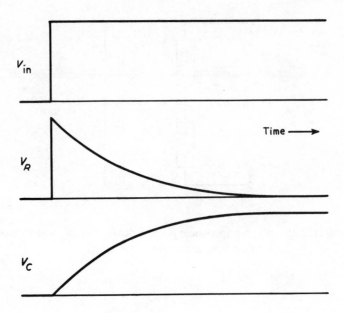

Fig. 11.4 Step input voltage (V_{in}) and the corresponding output voltages across the resistor (V_R) and capacitor (V_C) of Fig. 11.3.

Similar reasoning shows that if we take as the output voltage the potential difference between the ends of the resistor, then we shall allow the high frequencies to pass, but shall reject the low frequencies. We have, in this case, a *high-pass filter*.

The input voltage will not, of course, always be sinusoidal. Suppose we apply a sudden change of voltage, from one steady value to another steady value, as the input voltage. This is known as a step change in potential. In this case we find that the voltages measured across the capacitor (V_C) and the resistor (V_R) are as shown in Fig. 11.4. They change exponentially with time and their sum at all instants of time equals the input voltage. The reasons for the responses is as follows. Before the application of the step, the capacitor is charged up to a steady value, and no current can flow round the circuit. The voltage difference between the plates of the condenser equals the input voltage. Immediately on applying the step the condenser will begin to charge up or discharge, since its potential is now incorrect. This process of changing charge results in a current which flows through the resistor. The current will gradually decrease as the condenser charge nears equilibrium for the new potential. Since the potential across the resistor equals the current times the magnitude of the resistance it follows that the voltage V_r will be initially very high and will then decrease. We can investigate the responses mathematically. It is easiest if we investigate the response when V_{in} is suddenly switched on from being zero to being some potential V_{in}. From the previous chapter we know the potential difference across a resistance and a capacitance.

$$V_R = IR \tag{11.2a}$$

$$C\frac{dV_c}{dt} = I \tag{11.2b}$$

We also know two further things about our circuit. The first is that the current is the same at all points around it, so that the value of I in the above equation is the same in each. We also know that the sum of the voltages equals in input voltage,

$$V_{in} = V_R + V_C \tag{11.3}$$

Since the current in the above expressions is identical we have

$$C\frac{dV_c}{dt} = \frac{V_R}{R}$$

Substitute for V_R from equation 11.2

$$C\frac{dV_c}{dt} = \frac{V_{in} - V_C}{R}$$

i.e.

$$\frac{dV_c}{dt} = \frac{V_{in} - V_C}{RC} \tag{11.4}$$

This is a simple first order differential equation, and has a solution of the form

$$V_c = A + Be^{-\alpha t} \tag{11.5}$$

giving

$$Be^{-\alpha t} = V_c - A \tag{11.6}$$

in which A, B and α are constants that we still have to determine.

Differentiate equation 11.5, and substitute equation 11.6

$$\frac{dV_c}{dt} = -\alpha Be^{-\alpha t}$$

$$= -\alpha(V_c - A) = \alpha(A - V_c)$$

Comparing this with equation 11.4 above shows us that $\alpha = 1/RC$ and that $A = V_{in}$.

Thus

$$V_c = V_{in} + Be^{-t/RC}$$

In order to give a value to the constant B we have to make use of the fact that the condenser is initially discharged, and thus when $t = 0$ we know that $V_c = 0$. This gives a value for B of V_{in}, and our answer is

$$V_c = V_{in}(1 - e^{-t/RC}) \tag{11.7}$$

This is the mathematical equation of the expression drawn in the figure. RC is known as the *time constant*.

Substituting this expression into equation 11.3 gives

$$V_R = V_{in} e^{-t/RC} \qquad\qquad (11.8)$$

11.3   Measurement of resistance. The Wheatstone bridge

It is often necessary to measure either the absolute value of, or changes in, resistance. (For example, many transducers are resistors whose value is altered by the input variable being measured.) In principle this can be done by passing a known current through the resistor and measuring the potential drop across it. This requires a constant current source which is costly, and the apparatus will be liable to drift, giving erroneous measurements.

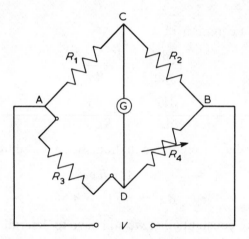

Fig. 11.5 Wheatstone Bridge circuit. R_3 is the unknown and R_4 the variable resistor.

The normal way of making such measurements is with the Wheatstone bridge circuit (Fig. 11.5). In this circuit the resistors in two of the arms are fixed (R_1 and R_2), the third is the resistor whose value is required and the fourth arm is a variable resistor. It is usual to make $R_1 = R_2$. An input voltage is applies across two opposite corners of the bridge and the output voltage is measured across the other two corners as shown. The output can be measured with a galvanometer or with a cathode ray oscilloscope. The procedure is to vary the magnitude of the variable resistor until there is no output across the output terminals. When this is done we know that the potential at point C (V_c) equals the potential at point D (V_D). It is also the case that no current flows across the output terminals, meaning that the current through R_3 ($= I_3$) equals the current through R_4, and the current through R_1 ($= I_1$) equals the current through R_2. Thus using Ohm's law we have

$$I_3 R_3 = I_1 R_1 \qquad \text{giving} \qquad \frac{I_3}{I_1} = \frac{R_1}{R_3}$$

and $\qquad\qquad I_3 R_4 = I_1 R_2 \qquad \text{giving} \qquad \dfrac{I_3}{I_1} = \dfrac{R_2}{R_4}$

Thus $\qquad\qquad\qquad\qquad R_1/R_3 = R_2/R_4$

We can rewrite this as

$$\frac{R_4}{R_3} = \frac{R_2}{R_1}$$

Since R_2 and R_1 are fixed resistors we can see that by varying the value of R_3 until balance is reached we can get an absolute measure or our variable R_4. The position of balance is not affected by the value of the potential applied across the input, and the device is therefore extremely stable.

The *sensitivity* of the apparatus *is* dependent upon the applied potential. By sensitivity we mean the smallest changes of resistance that we can distinguish. The ability to do this depends upon how much shift from zero output is obtained by a small change in resistance. The magnitude of this shift is directly proportional to the input potential. Thus for great sensitivity it is desirable to use large input voltages. However, the larger the input voltage the larger the currents passing round the circuit. and the greater the heating effect (power lost) in each resistor. Generally the variable resistor will have a maximum permissible power dissipation, and this will set a limit to the sensitivity.

Often it is only necessary to measure small changes in the variable resistance, since the parameter being measured is going through fairly small fluctuations, and it is the measurement of these that is required. In this case it is not necessary to continually vary R_3 in order to get balance (this would make rapid measurements impossible). Instead, provided the change in resistance R is small compared to R_4 (say not greater than about $1/10$ of R_4), then the output obtained will be directly proportional to R. The small fluctuations can thus be measured simply as the output voltage across the output terminals CD.

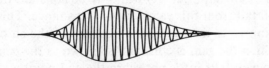

Fig. 11.6 Output waveform when input is sinusoidal for a small fluctuation in the variable resistor.

The input potential can be either constant (as from a battery) or variable. A sinusoidal voltage is often used. In this case the output will also be sinusoidal. Balance

is achieved as before by making the amplitude of the output voltage zero. A slight
imbalance will be sinusoidal, and the magnitude of the imbalance will be measured
as the amplitude of the output waveform. This is illustrated schematically in Fig. 11.6.
Notice that without making further measurements it is not possible to distinguish
a resistance increase from a resistance decrease in this case.

11.4   Voltage sources

A characteristic of any source of voltage, whether it is a battery, a power supply
or a membrane separating two solutions as in nerve and muscle, is that there is,
associated with that voltage, an internal impedance in series with it. Thus we should
draw any voltage source as shown in Fig. 11.7. The importance of this will become
obvious in the section on amplifiers.

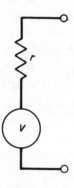

Fig. 11.7

11.5   Amplifiers

All too often, the signal being recorded is not of a suitable size to be measured by
the recording apparatus, and some form of amplification is required. Thus generally,
between the transducer and the recorder one has an amplifier.

The perfect amplifier is one which will amplify a signal of any frequency by the
same factor, and which introduces no other signal into the output.

In practice the gain of an amplifier (A) will only be constant over a certain range
of frequencies and will fall from this value outside this range. This dependence of
gain on frequency is known as the *frequency response* of the amplifier, and the range
of frequencies over which the gain is constant is known as the *bandwidth*. (Since
the point at which the gain falls off is not sharp the end point is defined as the
frequency at which the gain falls an arbitrary amount below the level of constant
gain – this amount is 0·86.)

Normally an amplifier is built up with a number of different stages (usually one
valve or transistor per stage) each of which amplifies the signal from the previous
stage. These stages can be connected together in two ways, with different effects to

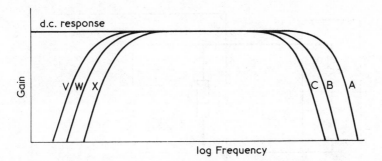

Fig. 11.8 Frequency response curves for d.c. and a.c. coupled amplifiers. Many amplifiers have controls for changing the frequency response. The effect of the high-frequency control is to shift the response curve from A to B to C as the frequency of 'cut-of' is lowered. The effect of changing the low-frequency control is to shift the curve from V to W to X. This is achieved with filters made from R–C circuits.

the response of the amplifier to d.c. (i.e. constant or *direct current*) signals. The most convenient way is to connect the stages by means of a condenser – when the amplifier is said to be *a.c. coupled* (a.c. stands for *alternating current*). In this case the amplifier gives no permanent deflection to a d.c. signal (although there will be a momentary kick as the signal is connected). Such an amplifier is very useful when it is required to observe a small varying output which is superimposed on a large permanent bias.

The stages of an amplifier can also be coupled, by means of resistors, to amplify constant signals. It is then said to be *d.c. coupled*. Such amplifiers are more difficult to build, and are therefore more expensive. Also it is a difficult problem to obtain an amplifier with no 'drift' of output signal with a constant input.

Another parameter of amplifiers which can lead to trouble is their input impedance (Z_{in}). This is the impedance that can be measured between the input terminals of the amplifier (when it is switched on). The amplifier only amplifies the voltage across this input impedance. This means that the voltage gain of the amplifier depends upon the value of the internal impedance of the voltage source. The circuit diagram for the input to an amplifier, including the voltage source whose voltage is being amplified is shown in Fig. 11.9.

This diagram is identical in principle to Fig. 11.2 from which we see that the voltage that will in fact be amplified (V_{in}) is:

$$V_{in} = \frac{Z_{in}}{r + Z_{in}} V_R \qquad (11.9)$$

Thus only provided the input impedance of the amplifier is much greater than the impedance of the source voltage, will the voltage being amplified be the source voltage.

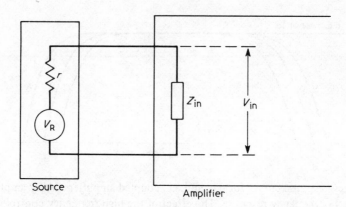

Fig. 11.9 Circuit diagram of input side of amplifier connected to a source voltage.

A typical value for the input amplifier of most oscilloscopes is 1–10 Megaohm.

An important application of this idea, for which the source impedance is extremely high is in the measurement of membrane potentials using microelectrodes. The apparatus is shown in Fig. 11.10.

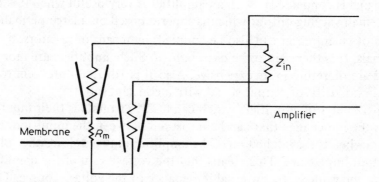

Fig. 11.10 Use of microelectrodes to measure membrane potential. Resistors are drawn within the electrodes to signify their resistance, and likewise for the membrane.

The source impedance in this case is equal to the resistance of each microelectrode in series plus that of the membrane. The resistance of a good microelectrode will be in the range 20–100 Megaohm, so in a normal experiment we would expect that the source impedance will be at least 40 Megaohm and possibly much higher. In this case the input voltage to a normal amplifier with an input impedance of about 1 Megaohm will be much less than that of the membrane potential. Worse still, the exact value will be very heavily dependent upon the microelectrode resistance. In any experiment

an experimenter may get through several microelectrodes, all of which will have
different resistances. Thus there will not only be a loss of voltage, but this loss will
be different for each microelectrode. This is obviously unsatisfactory. To overcome
the difficulty, special amplifiers are used with very high input impedances. In the
past these have been what are known as *cathode followers*, with an input impedance
of 100 to 1000 Megaohm. Transistor amplifiers with high input impedances are now
available. These use field-effect transistors (FETs) as their input stage, and can have
input impedances of 10^{12} Ohm.

The *output impedance* of an amplifier is also important. For voltage amplifiers,
rather than power amplifiers, it is important that the output impedance (which can
be thought of as the internal impedance of the output source of voltage) is small. For
transfer of power the optimum condition is that the output impedance of the
amplifier shall be equal to the impedance of the system to which power is being
transferred. Thus to drive a 15 ohm loudspeaker it is best if the output impedance
of the final amplification system is 15 ohm.

The full circuit diagram of an amplifier necessary for an understanding of its
applications is as shown in Fig. 11.11. The amplification factor (A) is frequency
dependent as shown in Fig. 11.8.

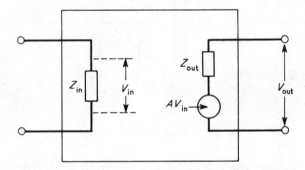

Fig. 11.11 Operation of an amplifier with gain A.

Usually, one of the input terminals of an amplifier is connected directly to earth,
and the signal on the other terminal amplified. However in many biological applications it is useful to have both input terminals identical, and the difference between
the two input signals obtained as output. This is especially useful for balancing out
interference which is common to both electrodes, say, when recording nerve inputs.
In this case a *differential amplifier* is used. This in effect has two input stages — one
for each electrode. It has three inputs. These are the two electrode inputs and an
earth connection.

11.6 Recording devices

The most versatile recording apparatus is the cathode ray oscilloscope (CRO).

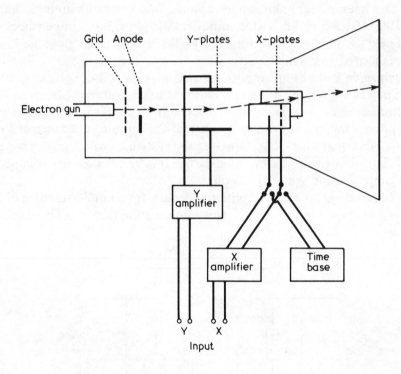

Fig. 11.12 Diagram of the CRO.

An oscilloscope consists of an electron gun which 'fires' electrons at a fluorescent screen. Since electrons travel only very short distances in air, the whole pathway must be evacuated. This evacuated volume is the oscilloscope *tube.* The electrons are produced by heating a filament by means of an electric current, and accelerated towards the fluorescent screen by an *anode* maintained at a high potential. The brightness of this beam is controlled by a grid whose potential dictates the number of electrons leaving the vicinity of the cathode. There is also a focusing device which can be considered as being a part of the gun, and this produces a *'focus'* control, and also on some oscilloscopes an *'astigmatism'* control.

We saw in the previous chapter that a moving charged particle can be deflected by either an electrostatic or an electromagnetic field. Most CROs use electrostatic fields to deflect the electron beam. This is produced by two *'deflection plates'* one either side of the beam. A difference of potential (voltage) between the two plates creates an electrostatic field and so causes the deflection. In an oscilloscope the beam can be deflected up and down or from side to side. One pair of plates (Y-plates) is required for the up-and-down movement, and another pair for the side-to-side movement (X plates).

Several tens of volts are required to cause a deflection of the electron beam from one side of the screen to the other. Thus an amplifier is nearly always required, and all oscilloscopes have an amplifier connected to the Y-plates (most research oscilloscopes also have amplifiers connected to the X-plates — in class oscilloscopes this is often rather rudimentary). This amplifier will have a *gain* control which enables the amplification of the amplifier to be set to different values.

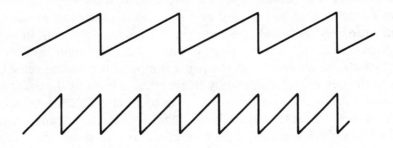

Fig. 11.13 Sawtooth waveforms of timebase.

The oscilloscope is normally used to provide a graphical display of a voltage waveform (which is fed into the Y-amplifier) against time. To enable this to happen a unit known as the *time-base* is connected to the X-plates. This unit produces what is known as a sawtooth waveform. The effect of this is to cause the electron beam to move with a uniform velocity from the left to the right of the screen, the velocity being determined by the rate of rise of potential in the rising phase. The lower waveform in Fig. 11.13 will make the beam move twice as fast as the upper one. When the maximum voltage is reached, the beam is on the RHS of the screen. The voltage is then lowered again to its initial value extremely rapidly (in class oscilloscopes in a few nanoseconds — see Table 2.1 for the meaning of 'nano'). The beam is returned to the LHS of the screen virtually instantaneously. The time-base can be made to give this waveform repititively, as in the diagram, in which case the beam is continuously traversing the screen, or it can be controlled to start only when the voltage on the Y-plates reaches some certain value. This is known as *triggering* the time base and is a very useful feature which enables synchronization of the experimental conditions with the traces on the oscilloscope screen.

Although you will normally use your oscilloscope with the time-base, you need not do so; in which case you will use it to feed in two different signals into the X and Y-plates, and you will use it as an X—Y plotter, producing a graph of one waveform against the other. This might for example be a graph of muscle tension against muscle length.

Permanent records of the oscilloscope records can be obtained by photographing the waveforms produced. This can be done by two methods. The most obvious is to simply photograph one sweep of the beam with the time-base connected. The other, which is used when recordings of events which take a long time are required (such as the nervous activity from a sense organ or other cell), is to stop the time base and to use a camera in which the film is moved. This type of camera uses rolls of film (often 25 or 100 feet at a time of 35 mm film) and has a control for various paper speeds.

The disadvantage of this method of producing permanent records is the trouble necessary to produce the records. A more convenient method which is useful for events which do not change too rapidly with time (which therefore excludes nerve impulses unfortunately) is to use a penrecorder. This is an instrument in which a pen moves from side to side across a piece of paper, the movement being controlled by the input voltage to the machine. The paper is moved at constant velocity at right angles to the pen movement, and by this method a trace of the event against time is produced. Penrecorders are made with up to 12 channels, so that simultaneous recording of up to 12 variables can easily be made. This is another advantage they have over the CRO which is usually restricted to about four simultaneous records, and often to only two.

One difficulty with neurophysiology is that to be sure of obtaining the response on permanent record a great deal of recording is required. This can be very costly in terms of paper for the camera of penrecorder. A way round this is to record experimental data on magnetic tape with a multicahnnel taperecorder, and to make visual records later. Normal hi-fi taperecorders cannot record events which change slowly with time (below about 30 Hz) but taperecorders are manufactured in which the input signal is first converted into an electrical signal whose frequency is proportional to the amplitude of the input. Thus an input of zero volts might give a frequency of say 20 kHz, + 1 volt a frequency of say 30 kHz and − 1 volt a frequency of say 10 kHz. The data can be stored on the tape in this form, and then played back through a unit which converts the frequencies back into voltages. This is known as *frequency modulation* and such taperecorders are known as F.M. recorders. The input signal cannot be faster than the lowest frequency used for storage, and such machines have a maximum frequency response about 2−4 kHz.

Another drawback of oscilloscopes is that the width of the trace produced on the screen is quite wide, and this restricts the accuracy obtainable. Often this is not important, but if great accuracy is required then some other form of recording the data must be found. Nowadays this is generally done with digital machines of one kind or another. Slow events can be recorded with digital voltmeters which produce a numerical display of the input voltage. Faster records can be recorded on a digital oscilloscope, or fed directly into a digital computer. These two machines both make use of *memory store*. The way this works is not important. The effect is that the machine has a 'memory' which is made up of many units (known as 'words'). Each

word can store just one number with high accuracy. A digital oscilloscope often
has about 1000 such words. The input waveform feeds its potential sequentially
into these words. Thus if the signal lasts one second, the average voltage of the first
millisecond will be stored in the first word, that of the second in the next word and
so on. This data can then be displayed either visually or else obtained as a set of
numbers which can then be analyzed. The digital computer can be made to do the
same, but it has the extra facility of being able to treat the data mathematically, so
that the data can be analyzed automatically. This kind of apparatus enables both
the attainment of high accuracy, and also the treatment of a large amount of data,
which would otherwise take a prohibitively long time.

11.7 Transducers

A transducer is a piece of apparatus which either converts inputs of a non-electrical
nature into an electrical signal, or converts electrical inputs into some other form,
for example movement of an arm. This section will deal with the former. Numerous
types of transducer are used. Those given below are a selection.

(a) Transducers depending upon a change in resistance. One model of this type was
dealt with in the section on semiconductors — the heat sensitive thermistor. Semiconductor and wire strain gauges are also transducers in this class, and can be
used for measuring tension. All such are generally used in a Wheatstone bridge
circuit.

(b) Transducers depending upon a change in inductance or capacitance. In any
capacitor, the value of the capacitance is dependent upon the distance apart of the
plates. Varying this distance therefore varies the value of the capacitance. Similarly,
movement of a core within an inductance coil changes the value of the inductance.
Both can readily be used as length transducers. Values of capacitance and inductance
are most conveniently measured by making them part of a resonant circuit, and
measuring the change obtained in this circuit.

(c) Transducers depending upon a change in light intensity. Various methods of
measuring light intensity are known — photocells, semiconductor photoresistors
and phototransistors being the most commonly used. Apart from the obvious use
of measuring the intensity of light, such devices are often used as the detector in
length transducers which work on the principle of moving a vane through a light
beam, thereby changing the intensity of light reaching the light detector. Solar
cells, which give a voltage output proportional to the intensity of light falling on
them can also be used.

(d) Valve transducers. A commonly used one of these is the tension transducer
with the trade number RCA 5734. In this valve the anode is able to move, and has
a deflection which is proportional to the tension applied. This causes a change in

the anode current which can be measured.

(e) Temperature transducers. The semiconductor thermistors have already been mentioned. The second commonly used form of temperature measuring device is the thermocouple giving an output voltage proportional to temperature difference between two junctions. This device can be made more sensitive by using several in series, when it is known as a thermopile.

11.8　Noise and interference

These are both unwanted signals which are introduced by the recording apparatus.

(a) Thermal noise. An electron in a conductor has thermal energy (of magnitude $3\,kT/2$ – where k is Boltzmann's constant and T is the absolute temperature) and moves around with a mean kinetic energy equal to this thermal energy.

i.e.　　　　　　kinetic energy $=\ 1/2mv^2\ =\ $ thermal energy $=\ 3\,kT/2$

$$giving\ v\ =\ \sqrt{\left(\frac{3\,kT}{m}\right)} \tag{11.10}$$

A movement of electrons is an electric current, and this current moving in a resistance will create a voltage across that resistance. This is the origin of thermal noise. Since the movement of the electrons is random, the noise will be random, and the energy of the noise will be evenly distributed over the whole spectrum of frequencies. We can now understand the origin of all the terms in the expression for the thermal noise developed in a resistance R:

$$R.M.S.\ voltage\ due\ to\ noise\ =\ \sqrt{(4\,kTBR)} \tag{11.11}$$

where R.M.S. voltage stands for 'Root Mean Square' voltage and is just a convenient way of measuring a mean amplitude for the noise voltage, and B is the bandwidth (i.e. the frequency range covered in Hz – thus if the frequency range is from 10 to 10 000 Hz the bandwidth is 9 990 Hz).

One way of reducing the noise therefore is to restrict the bandwidth. You will find for example on many preamplifiers you will use that there are two controls for changing the bandwidth; one alters the low frequency end, and the other the high frequency end. Such controls simply introduce different filters into the circuit.

(b) Interference. This is the term given to pick-up unwanted electric and magnetic fields. The most common source of such interference is the 50 Hz mains, which gives rise to what is commonly called 'hum'.

(i) Electrostatic interference. This is caused by the presence of charged material in the vicinity of the apparatus. Leads to other equipment are a common source. In the previous chapter we saw that the electric field inside a closed metal box due to external charges is zero. The box should be earthed so that it remains at zero potential.

Thus the method of eliminating electrostatic interference is to screen the apparatus and leads. It will often not be necessary to place the entire apparatus inside a metal box, but will be sufficient to use screened leads. These are leads surrounded by a copper braiding, known as coaxial cable, because the inner load is surrounded coaxially by the outer braiding. The screening should be earthed. Another method of reducing electrostatic interference is in the use of differential amplifiers. the two leads from the preparation are taken to the two inputs of the differential amplifier. Since the pick-up will be approximately the same in the two wires most of it will be rejected by the amplifier.

(ii) Electromagnetic interference. This is caused by the flow of current in a conductor, and is picked up by loops in a circuit. A common loop, known as an earth-loop is formed by a piece of apparatus being earthed at two different points. This can give rise to induced current round this loop. Thus apparatus should be well earthed, but at only one point. Another source of loops is formed by the pair of leads leading from one apparatus to another or from a preparation to the input of the electrical apparatus. These are best eliminated by twisting the two leads together, in which case the small loops formed by each twist are of opposite sense for adjacent twists and cancel each other out. In extreme cases a piece of apparatus can be shielded by a screen made of material of high magnetic permeability (such as iron or μ-metal). Magnetic lines of flux tend to congregate in materials of high permeability) and thus the magnetic field in such a box is very low.

11.9 Questions

1. What will be the value of the thermal noise in a resistor of (a) 1000 Ω (b) 1 MΩ when the band width is 100 kHz.

2. A neurophysiologist uses two microelectrodes of resistance 20 MΩ to record membrane potentials, and connects them to an oscilloscope whose input impedance is 2 MΩ. By what factor are his measurements inaccurate?

12 Oscillations, Wave Motions and Sound

Before discussing oscillations and wave motions we must first understand the meaning of sinusoidal motion, and in particular the various parameters used to describe the properties of a variable which is changing sinusoidally with time.

12.1 Sinusoidal oscillations

Oscillations are a very common phenomenon, both biologically and non-biologically. An insect beats its wings in an oscillatory fashion, the heart causes the pressure in the vascular system to be oscillatory. Sound is produced by oscillations in air pressure, and light by oscillations in electromagnetic radiation. The list is endless.

What are the characteristics of oscillations? The main feature is that they are a repeating pattern of some variable. This means that the variable increases and decreases with time in a regular fashion. Fig. 12.1 demonstrates some oscillations. The waveform, meaning the shape of the repeating pattern is different in all these. Is there a 'simplest' form of oscillation that is fundamental to all oscillations? The answer to this is that there is, and it is known as *sinusoidal* oscillation.

The reason that the sine wave, as the waveform of sinusoidal oscillation is known, is the simplest to produce is that its production has only one requirement, that the acceleration of the oscillatory variable when it is displaced from its mean position be proportional to its displacement from that position and directed towards that position. Any other form of oscillation requires more rules for its production than this. Mathematically this can be written:

$$\frac{\mathrm{d}^2 x}{\mathrm{d}t^2} = -kx \tag{12.1}$$

We will return to this idea later. The simplest way of getting some idea of the shape of the sine wave is to consider one way of producing a sine wave. This is by means of circular motion, with a constant angular velocity ω. Imagine a rod rotating about one end as in Fig. 12.2. Drop the perpendicular from the end of this rod onto the x-axis in order to obtain the projection of the rod on the x-axis. Call the magnitude

174

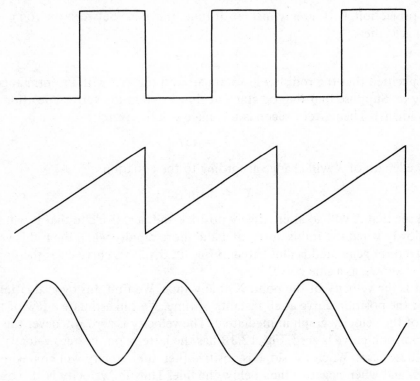

Fig. 12.1 Various oscillations. The bottom one is a sine wave.

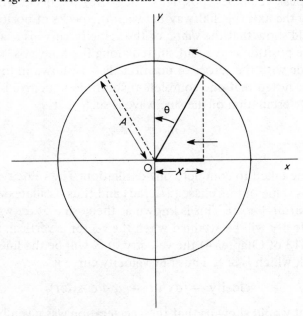

Fig. 12.2 Derivation of a sine curve from a constant angular velocity. The component of the radial arm in the x-direction traces a sine curve.

of this projection X. If, at any instant of time, the angle between the rod and the vertical is θ, then

$$X = A \sin\theta \qquad (12.2)$$

We specified that the rod should rotate around the axis with uniform angular velocity ω. Suppose that the rod starts at time $t = 0$ in the vertical position (i.e. $\theta = 0$ radians). Then after t seconds the angle will be given by

$$\theta = \omega t \qquad (12.3)$$

Thus the length of X will change according to the equation

$$X = A \sin(\omega t) \qquad (12.4)$$

We can see that X will go alternatively positive and negative and that it will move fairly slowly when the rod is horizontal and more rapidly when the rod is vertical. The waveform generated is illustrated in Fig. 12.3A. This curve gives the position of X, and is known as a sine curve.

What is the velocity of the point X at any time. We find this by evaluating the slope of the position curve at all instants of time. We can estimate a few of the points of the velocity graph immediately. The velocity is zero whenever the slope of the position curve is zero. Fig. 12.3B has the letter z on the zero velocity axis for all these cases. When the slope is positive then the velocity will be positive (above the axis) and when negative then below the line. Thus the velocity is also oscillatory. Furthermore the peak velocities (positive or negative) will occur when the position curve was crossing the axis (i.e. halfway between the peaks of position). A full investigation would show that the *shape* of the velocity curve is exactly the same as the *shape* of the position curve, but shifted along the time axis so that the peaks of velocity coincide with the zeroes of position. This is shown in Fig. 12.3C.

In Chapter 4 we noted that one complete revolution of a circle is 2π radians. Thus one complete oscillation of the sine wave is such that

$$\omega t_0 = 2\pi \qquad (12.5)$$

where t_0 is the time taken to complete one oscillation. The velocity waveform is one quarter of this value out of phase ($\pi/2$ rad) and thus oscillates with a waveform proportional to $\sin(\omega t + \pi/2)$. This is known as the cosine of ωt written $\cos(\omega t)$. The maximum velocity will be attained when the vector is vertical, at which time (using equation 4.13 of Chapter 4) the velocity of X will be the linear velocity of the end of the rod, which is ωA. Thus our velocity curve is

$$\text{velocity} = dx/dt = \omega A \cos(\omega t) \qquad (12.6)$$

A similar argument would show us that the acceleration was given by

$$\text{Acceleration} = -\omega^2 A \sin(\omega t) \qquad (12.7)$$

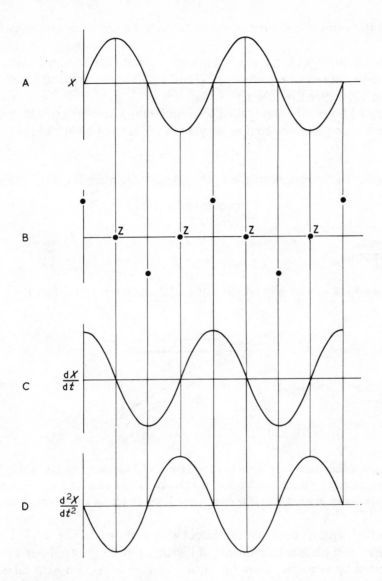

Fig. 12.3 A. The sine curve generated by the x-component of the arm of Fig. 12.3. B. The points of maximum and zero velocity of curve A. C. The full curve of the velocity of the parameter X. D. The acceleration curve of X.

Notice that $A \sin(\omega t)$ equals the position of X. Thus

$$\text{Acceleration} = d^2 X/dt^2 = -\omega^2 X \qquad (12.8)$$

ω is a constant angular velocity, and thus ω^2 is also a constant. In other words our sine curve has justified the statement made earlier that it is the curve produced when the acceleration towards the median point equals the displacement away from that point.

If you are familiar with calculus, then you will not need the long explanation given above to know that equations 12.6 and 12.7 are the first and second derivatives of equation 12.4. (See Chapter 1.)

Equation 12.5 gave the time taken to complete one revolution, which is just the time taken to complete one complete cycle of the sinusoidal motion as

$$\text{Period} = t_0 = 2\pi/\omega \tag{12.9}$$

The frequency of recurrence of the cycle is $1/t_0$ .1 cycle/sec is called 1 Hertz.

$$\text{Frequency} = \omega/2\pi \tag{12.10}$$

12.2 Damping

Let us have a look at a case in which sinusoidal motion is produced, the insect wing.

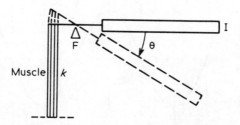

Fig. 12.4 A simplified insect wing of inertia I and free to rotate about the fulcrum F. The total elastic elements comprise the exoskeleton of the thorax and the muscles. For the example they are drawn as a single muscle of stiffness k, which acts as a Hookean spring.

A simplified representation of the insect wing is shown in Fig. 12.4. The wing can be considered as a plate with moment of inertia I free to rotate about a fulcrum F, and supported by a muscle, or by the cuticle of the thorax, whose total elasticity is contributed to by the thoracic muscles. We will simplify the action of the muscle by supposing that during flight it acts as a spring of elasticity k. How will the wing behave in this simplified model?

Suppose that the wing is depressed, as shown by the dotted lines, through an angle θ. The muscle will exert a restoring force proportional to its linear extension (x) of kx. This will exert a torque on the wing, which for small angles (as in all cases we are assuming that θ is sufficiently small that $\sin\theta \doteq \theta$) will be proportional to θ.

$$\text{torque} = k\theta \tag{12.11}$$

The effect of the torque will be to cause an acceleration of the wing (see page 40) of magnitude $I \, d^2\theta/dt^2$ and in a direction which will tend to reduce θ.

Thus
$$\text{torque} = -\, I \frac{d^2\theta}{dt^2} = k\theta \qquad (12.12)$$

This is exactly the same form as equations 12.1 and 12.8 and shows that the motion of the wing if released from this point will be

$$\theta = A \sin(\omega t) \qquad (12.13a)$$

where
$$\omega = \sqrt{k/I} \qquad (12.13b)$$

$$\text{frequency} = \frac{1}{2\pi} \sqrt{\frac{k}{I}} \qquad (12.13c)$$

A is the amplitude of the movement.

In many orders of insects (diptera, hymenoptera, coleoptera and some hemiptera) the frequency at which the wings beat is determined by these considerations. Thus cutting part of the wings from the animal causes the increase in frequency of wing beat that is predicted from equation 12.13.

The story as presented is obviously too simple. The reason for this is that if the wing beats depended only on an elastic muscle then oscillations of the wing would quickly die away. For example, support a ruler over the edge of the bench, bend it down and quickly release it. The ruler will oscillate for several cycles but these will gradually get smaller and smaller and will die away to nothing. The cause of this is *damping forces*, in this case largely the viscous drag of the air. The general characteristic of the factors responsible for damping in any system is that there is a force which is dependent upon the *velocity* of the displacement.

We can rewrite equation 12.12 as

$$I \frac{d^2\theta}{dt^2} + k\theta = 0 \qquad (12.14)$$

The addition of a force proportional to the rate of change of θ gives

$$I \frac{d^2\theta}{dt^2} + b \frac{d\theta}{dt} + k\theta = 0 \qquad (12.15)$$

b is known as the viscosity of the system and the solution to this equation is a sine wave whose amplitude dies away exponentially to zero with a time constant $1/b$. If b is too large then no oscillations are seen, the damping has taken place within one cycle.

The flight muscles of the orders of insects listed above work because the property of the flight muscles is to exert a *negative* damping which removes the effect of the normal damping. Since the velocity dependent term is the property of a viscosity, this term is often known as the viscous damping, and the insect flight muscle is said to have negative viscosity.

12.3   Energy of an oscillation

Let us continue with the previous example of the insect wing. The oscillations were initiated by causing the wing to move from its equilibrium position (when $\theta = 0°$ there is no force acting on the wing, which is therefore in equilibrium). Taking the wing from its equilibrium position requires work, and the energy is stored in the elastic elements in the form of potential energy. The magnitude of the potential energy (from page 57) is $\frac{1}{2}k\theta^2$. The maximum deformation from equilibrium is the amplitude A of the ensuing oscillation, at which point the energy stored is $\frac{1}{2}kA^2$. The potential energy stored in the elastic elements will vary from this value to zero during the oscillation as the displacement varies between the extreme values and the equilibrium position. During the oscillation the wing will gain kinetic energy. The magnitude of the kinetic energy at any point will be $\frac{1}{2}I\omega_0^2$ (see page 40) where $\omega_0 = d\theta/dt$. Notice that ω and ω_0 are different. The total energy of the system at any point of the cycle will be the sum of the kinetic energy and the potential energy at that point. We know from equations 12.4 and 12.6 that

$$\theta = A\sin(\omega t) \quad \text{and} \quad \omega_0 = d\theta/dt = \omega A\cos(\omega t)$$

Thus the total energy

$$E = \tfrac{1}{2}k\theta^2 + \tfrac{1}{2}I\omega_0^2 = \tfrac{1}{2}kA^2\sin^2(\omega t) + \tfrac{1}{2}I\omega^2 A^2\cos^2(\omega t)$$

We know that $\omega^2 = k/I$ (equation 10.13b)

Therefore
$$E = \tfrac{1}{2}kA^2\sin^2(\omega t) + \tfrac{1}{2}I\frac{k}{I}A^2\cos^2(\omega t)$$

$$= \tfrac{1}{2}kA^2\{\sin^2(\omega t) + \cos^2(\omega t)\}$$

$$= \tfrac{1}{2}kA^2 \text{ since } \sin^2\beta + \cos^2\beta = 1 \text{ for any value of } \beta. \qquad (12.16)$$

Therefore the energy of a sinusoidal oscillation is given by $\frac{1}{2}kA^2$ where A is the amplitude and k is the stiffness of the restoring force. A^2 is known as the *intensity* of the oscillation, and is a term often used when specifying a wave motion.

12.4   Waves

The oscillations discussed in the preceding sections have the property that the motion is simply a function of the time, in that there is no propagation through space of the motion. The variable in the oscillation (the position of the insect wing in our example) may indeed be moving in space, but the second cycle traverses exactly the same path as the first; there is no progression in *space*, although there is in *time*.

The property of a wave motion that distinguishes it from a simple oscillation is that there is progression in space. Some examples will illustrate this. A stone thrown into a pond causes waves to be initiated which progress outwards forming steadily increasing circles in the pond. A long piece of string, if suddenly jerked will have a disturbance which passes along the string. Sound is a wave motion; a noise propagates through the air at a finite velocity. An action potential in a nerve is a wave of potential difference passing along the nerve.

A second property of a wave motion is that the disturbance which characterizes the wave (the ripples in the pond, the side-to-side motion in the string, the pressure changes in the air and the membrane potential are the disturbances in the three examples above) propagates through the medium (the water surface, the string, the air and the nerve) without giving to that medium any permanent displacement. Thus a piece of cork floating in the water is not moved sideways by the passage of the wave, just up and down as the wave passes. Sound does not propagate through the air because gas is transported 'carrying' the sound, but because the pressure change in one place causes pressure changes in adjacent places. There is a side-to-side movement of gas molecules, but after the sound has passed they are in the positions that they would have had, had there been no sound.

It is simplest to consider the progression of a sinusoidal wave. This is more general than might appear at first sight because any wave can be considered to be made up of a combination of sinusoidal waves of different frequencies and amplitudes.

The wave will progress through the medium with a certain velocity, c, and provided the medium does not change its properties this velocity will be constant. Suppose that at any instant of time we take a look at the disturbance caused by the wave at different points in space; see Fig. 12.5. A short time later the wave will have progressed and the disturbance/distance graph will now be the dashed curve.

The equation of the sinusoidal wave motion depicted in Fig. 12.5 is

$$\phi = A \sin \left(\frac{2\pi}{\lambda} (x - ct) \right) \tag{12.17}$$

λ is known as the wavelength, and c is the velocity. We can see that this makes sense by looking at the shape of the curve at one instant of time. Time $t = 0$ is the most convenient. At this time

$$\phi = A \sin \frac{2\pi x}{\lambda} \qquad \text{at } t = 0 \tag{12.18}$$

When $x = \lambda$ then $\phi = A \sin(2\pi)$ and we have completed one complete cycle. The

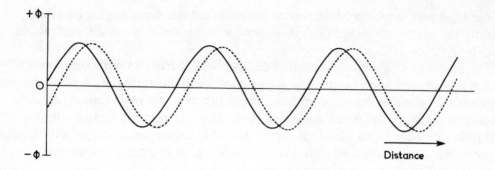

Fig. 12.5 A sinusoidal wave depicted at two times a short interval of time apart. The entire wave has moved along the distance axis.

wavelength can therefore be seen to represent the distance occupied by one cycle in space at an instant of time.

We can go through a similar argument for the curve of the disturbance at one place as a function of time. The easiest position to consider is $x = 0$, at which

$$\phi = A \sin(-2\pi ct/\lambda) \qquad \text{at } x = 0 \qquad (12.19)$$

When $t = \lambda/c$ then $\phi = A \sin(2\pi)$ and we have completed one cycle. This value of λ/c is known as the period t_0 of the wave, the inverse of t_0 being the frequency n.

$$t_0 = \lambda/c \qquad n = c/\lambda \qquad (12.20\text{a, b})$$

12.5   The general wave equation

This is the differential equation that specifies a wave motion. We shall only consider one dimensional wave.

It is
$$\frac{d^2\phi}{dx^2} = \frac{1}{c^2}\frac{d^2\phi}{dt^2} \qquad (12.21)$$

We can see that our sinusoidal wave $\phi = A \sin(2\pi/\lambda(x - ct))$ is one solution to this equation by differentiating this first with respect to x and then with respect to t. A more general solution is

$$\phi = f(x - ct) \qquad (12.22)$$

where f is any function.

12.6   Reflection, transmission and absorption

What happens when there is a boundary in the medium of the wave motion? This might be the edge of the pond or a piece of string, the interface between air and a solid, in fact any point at which the uniform properties of the medium change. Something has to happen to the energy of the waves reaching this boundary. This something can be one of three things; reflection, transmission or absorption. It can also be a mixture of the three. Let us illustrate this with sound. Suppose you shout at a brick wall. You will be able to hear an echo (provided you are far enough away that you can distinguish the echo from your original shout!). This is due to the sound waves being reflected from the wall. A person behind the wall will also hear your shout, since it has been transmitted through the wall. Both the echo and the sound behind the wall will be quieter than your original shout. In fact the total energy of the echo and the transmitted sound will be less than that of your original shout because some of the energy will have been absorbed by the wall. This is absorption and results in a heating effect on the wall. The shout imparted kinetic energy to the molecules of the wall, some of which remained and was not used to re-excite the air on the other side. Exactly how much energy ended up in each of the three forms depends upon the properties of the wall. Cover it with felt and most of the energy will be absorbed. Make it very rigid and thick and you will increase the reflected energy and reduce that transmitted.

One particular instance of importance (in understanding the properties of sound producing and hearing organs) is the situation when a piece of string or fibre or a sheet of membrane is held rigidly at its extremities. In this case nearly all the wave motion set up is reflected from the ends, and this results in standing waves, discussed in the next section.

More complex situations can arise than sharply defined boundaries. One very important biological example is the inner ear. Here the properties of the medium (the basilar membrane) change uniformly down the length of the membrane. This results in a wave motion whose amplitude changes as it progresses down the membrane.

You will meet at least four adjectives describing types of wave motion.

1. Progressive, or more usually, *travelling* waves. This is the type of wave we have so far been describing.
2. *Standing* waves. This is the type of wave motion that is set up in a piano string. If the extent of the medium over which the wave motion can pass is limited, and if reflection of the wave can occur at the boundaries then standing waves will be set up. The piano string is held with its ends at a fixed position so that they are unable to move. A deflection of the string will set up vibrations in the string which for a given length of string under a given tension have a precise frequency and waveform. These are known as standing waves.

In many ways standing waves resemble simple oscillations; in particular there is a continual transfer of energy during a single cycle between that stored as potential energy and that in the form of kinetic energy. Although the total energy remains

constant, the kinetic energy and the potential energy both vary during the cycle.
This is not the case in a travelling wave in which the kinetic energy and the potential
energy of the wave remain constant and equal to each other.
3. *Transverse* waves. The disturbance that gives rise to the wave motion can either
be an up-and-down motion in that the direction of the disturbance is perpendicular
to the direction in which the wave is travelling, or it can be side-to-side so that the
disturbance is in the same direction as that of the resulting wave. The ripples on
the water are an up-and-down movement. The water surface moves vertically, whereas
the wave moves horizontally. This is known as a transverse wave.
4. *Longitudinal* waves. These are the waves in which the disturbance is in the same
direction as the ensuing wave. Sound is the example that you will meet with this type
of disturbance.

12.7 Wave packets

The wave motion that traverses a medium need not be continuous, although of
course it can be. A continuous sound wave would be, say, the continuous production of a note from a loadspeaker produced by an oscillator. More often however the sound will be much shorter than this, at the extreme being a bang. In this
case the sound progresses through the air in the form of a local 'packet' of
oscillations as shown in Fig. 12.6.

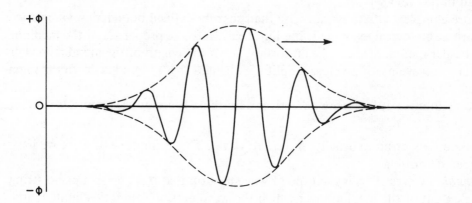

Fig. 12.6 A wave packet.

There will be a set of oscillations whose amplitude varies from zero at one end to
zero at the other, with some 'group' pattern in between. This can readily be seen by
throwing a stone into a lake. A set of ripples will spread out from the centre; there
will be about ten or so peaks present at any one time, and this group of oscillations
will move across the surface of the water. Notice that the overall velocity of the group

of waves, known as the *group velocity* can be different from the velocity of the
oscillations within the group, known as the *wave velocity*.

12.8 Resonance

An oscillator or a source of standing waves has a certain frequency at which it will
produce oscillations. This is known as the *resonant frequency*.For the insect wing
this was given by equation 12.13b.

How can waves be set up in such a system? Let us consider the case of a simple
oscillating system. We will continue with the insect wing — the arguments are
applicable to any system that is described by the same equations. One way of
starting oscillations is to take the system away from equilibrium and release it. If
the system is damped the oscillations die away. How can we maintain oscillations
in the system. What happens if we apply a periodic force and in particular a sinusoidal
force. If the frequency of the applied force is very different from the resonant
frequency of the system, then the disturbance obtained will be very small. In our
example we can apply a force to the system by moving the lower end of the muscle
in Fig. 12.4 up and down cyclically. Away from the frequency of natural oscillations
in the system this will produce only a very small movement of the wing. At very
low frequencies this movement will equal that applied. However, if the frequency
of the applied force is very close to the natural oscillation frequency then the wing
motion will be very large. The graph of the amplitude of movement of the wing as a
function of the applied frequency of force of constant amplitude applied to the
system is shown in Fig. 12.7. We say that the system *resonates* as its frequency of
natural oscillation. This effect can be seen on the piano. Hold any note down
without striking it. This removes the damping pad from the string. Strike a note which
is one octave lower briefly. You will hear the higher note resonating for as long as
you hold it down. Notice that at resonance the energy that the system has achieved
is much greater than at other frequencies.

Resonance is important from the point of view of sound reception by animals.
For example many insects have organs to pick up sound waves which are essentially
hollow cavities (made from trachea), one face of which consists of a taut membrane.
This structure is known as a tympanum, and has the property that it resonates at a
particular frequency; a kettle drum is a similar type of structure with exactly the
same properties. The membrane is itself able to resonate, and because the space
behind it is enclosed it sets up oscillations in the air in this space. If the frequency
of the incoming sound is very different to the resonant frequency of the tympanum,
then the movement induced in the membrane will be very small, and the sense cells
which are capable of detecting movement in the membrane will not be excited. When
the incoming sound is at the resonant frequency, then the membrane will perform
large movements, and the sound will be detected because these movements will be
sufficient to excite the sense cells.

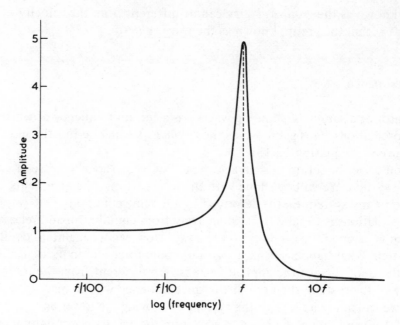

Fig. 12.7 The way in which the amplitude of the wing of the system of Fig. 12.4 moves if the lower end of the muscle is moved through a constant amplitude sinusoidally at different frequencies, shown as a function of the frequency for a small amount of viscous damping. Amplitude of movement = 1 on scale.

Another important type of resonance is found in the ability of electromagnetic radiation, for example light, to excite transfers between the various energy levels at the molecular level of matter. This is described in more detail in Chapter 14.

12.9   Change in intensity with distance

The energy emitted from a source of waves can be transmitted in one, two or three dimensions depending upon the type of wave being propagated. Thus waves travelling down a nerve are one-dimensional, ripples in a pond are two-dimensional and sound and light are three-dimensional. In the absence of frictional effects the total energy of the entire wave will be constant.

The effect of this is that the energy waves reaching a point will vary with distance in different ways according to the type of wave.

1. One-dimensional waves will have a constant intensity.
2. Two-dimensional waves will obey: intensity $\propto 1/r$.
3. Three-dimensional waves will obey: intensity $\propto 1/r^2$.

in which r is the distance between the point of origination of the waves and the point at which the intensity is measured.

Thus sound and light waves will have an intensity which varies as the inverse square of the distance away from the point of origination.

12.10 Sound

Sound is transmitted through a compressible medium (gas or liquid) or an elastic medium (solid) as longitudinal pressure waves. This means that the disturbance can be thought of as either fluctuations in pressure or as the side-to-side movement of particles of the fluid or solid which are causing that change in pressure.

The *velocity* of sound in the medium is dependent upon the relationship between pressure (p) and density (ρ) in that medium. That this is the situation can be understood in that the transmission of sound requires the movement of particles of the medium; the forces are due to the pressure gradients, and the ensuing accelerations are dependent upon the density. The pressure gradients can themselves be described in terms of density gradients.

The wave equation for such longitudinal waves in a fluid is:

$$\frac{d^2\phi}{dx^2} = \frac{d\rho}{dp}\frac{d^2\phi}{dt^2} \qquad (12.23)$$

where ϕ is the displacement of the fluid particles. By comparison with equation 12.21 we see that the velocity of sound is

$$c = \sqrt{(dp/d\rho)} \qquad (12.24)$$

In the case of a fluid of Bulk Modulus K and mean density ρ

$$c = \sqrt{(K/\rho)} \qquad (12.25)$$

In the case of a solid of Young's modulus Y and density ρ

$$c = \sqrt{(Y/\rho)} \qquad (12.26)$$

Another important parameter is known as the *characteristic impedance* of the medium. This is of particular importance when considering the transmission of sound across an interface – a common biological phenomenon that occurs between the pressure waves of the medium of the animal and the ear. The definition of impedance is closely analogous to that in electricity. In this case it is defined as the ratio:

$$\text{Characteristic impedance } Z = \frac{\text{amplitude of pressure variation}}{\text{amplitude of velocity variation}} \qquad (12.27)$$

Its value is given by:

TABLE 12.1

	Density kg/m^3	K N/m^2	c m/s	Z kg/m^2/s
Air	1·2	1·4 × 10^5	343	412
Water	10^3	2·2 × 10^9	1,470	1·47 × 10^6

$$Z = \rho c \qquad (12.28a)$$

$$= \sqrt{(\rho K)} \text{ for fluids} \qquad (12.28b)$$

At a boundary between two media of characteristic impedances Z_1 and Z_2, the fraction of the energy reflected at the interface is

$$\text{Fraction reflected} = \left(\frac{Z_1 - Z_2}{Z_1 + Z_2}\right)^2 \qquad (12.29a)$$

$$\text{Fraction transmitted} = 1 - \text{fraction reflected} \qquad (12.29b)$$

The values of the relevant parameters needed in air and water are shown in Table 12.1. The values of the impedances used in equation 12.29 must refer to the same area of cross section.

12.11   Sound detection

There are basically two methods of detecting sound. A. To detect the *pressure* changes in the medium. This requires a detector which is enclosed as in Fig. 12.8A. Such detectors are *non-directional.*

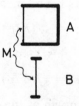

Fig. 12.8 Sound detectors. A. Pressure detector. B. Displacement detector.

B. To detect the *displacement* of the particles of the medium. This requires an open detector as in Fig. 12.8B, and such detectors are *directional*, and known as displacement of velocity detectors.

Another problem in sound detection concerns impedance matching. This is particularly acute in vertebrate ears, in which the inner ear is fluid filled, whereas the outer ear is air filled. The normal transmission across an air/water interface is extremely poor. Fitting the values of the characteristic impedances of air and water into equation 12.29b give a transmission of energy of

$$1 = \frac{1\,470\,000 - 412}{1\,470\,000 + 412}^{2} = 1 \cdot 12 \times 10^{-3}$$

This is an insignificant proportion. To overcome this the ear has developed an impedance matching set of devices. These are two-fold. In the first place the area of reception of the waves in the air is about 20 times that of the area of the bone which transmits the movement to the cochlea. Further there is a lever system in the bones of the middle ear which gives a displacement reduction of between 2 and 3 times. These both act to reduce the effective impedance of the cochlea.

In the ear the situation is also improved by virtue of the fact that the compliant membrane in the round window allows bulk movement of the fluid in the ear; the effect of this is to lower the effective characteristic impedance of the fluid. The very high value of 1 470 000 no longer holds for the situation in the ear, but must be reduced by a factor of about 10.

12.12 Doppler effect

Both light and sound move through their respective media at a velocity which is independent of the velocity with which the source is moving. The effect of this is to make the frequency f (and the wavelength) of the sound or light observed by an experimenter depend upon the relative velocity between the source and the observer.

Fig. 12.9

Consider the situation in which both the source and the observer are stationary. A frequency f_s Hz emitted by the source will be observed as that frequency f_s by the observer. In one second the wave train will be c metre long and contain f_s oscillations, each oscillation occupying $\lambda_s = c/f_s$ metres. Suppose now the source moves at a constant velocity v m/s towards the observer. After one second the first oscillation will be c metre from the original position of the source, but the last oscillation in this second will this time be v metre from this original position. The apparent wavelength has been reduced to $(c - v)/c$ of its value when the source was stationary.

i.e.
$$\lambda_m = \frac{c-v}{c}\lambda_s$$

$$f_m = c/\lambda_m = \frac{c}{c-v}\,\frac{c}{\lambda_s} = \frac{c}{c-v}\,f_s$$

13 Light

According to the type of situation being considered it is most straightforward to emphasize different (and to some extent at first sight seemingly contradictory) properties of light. Thus when discussing the *interaction* of light with matter it is helpful to consider light in the form of particles or photons; when discussing the *propagation* of light we shall sometimes think in terms of a wave motion emanating from a source, and at other times in terms of a stream of particles (photons) travelling in straight lines – a light ray. Quantum mechanics is the subject which unifies all these simplifying viewpoints, but is well beyond the scope of this book.

Light is one form of electromagnetic radiation. There are many other names given to different forms of electromagnetic radiation (radio-waves, infra-red, X-rays, etc.); the difference between these various forms is simply one of their wavelength or frequency. Fig. 13.1 indicates what these are. The velocity of light is different in different materials; it is greatest in a vacuum, and in a vacuum the velocity of all frequencies of electromagnetic radiation is the same. The ratio of the velocity of light *in vacuo* to that in a given material is known as the refractive index of that material. The relationship between the wavelength (λ) and velocity (c) of electromagnetic radiation is

$$\lambda = c/\nu \tag{13.1}$$

where ν is the frequency. When light passes from one material to another its frequency remains the same, but its wavelength and velocity change.

Under most circumstances you will find it best to think of light and other forms of electromagnetic radiation as a wave motion. As the name 'electromagnetic' implies, the disturbance responsible for the radiation is a joint electric and magnetic field. The oscillations of the electric and magnetic fields are sinusoidal and perpendicular to the direction of travel of the wave, and perpendicular to each other (Fig. 13.2). From our point of view we need simply consider one or other of the two forms of disturbance – we shall choose the electric field. In this case we can describe light both in terms of its direction of travel, and in terms of the direction of the oscillation of its electric field. This direction of oscillation of the electric field is known as the direction of *polarization* of the light.

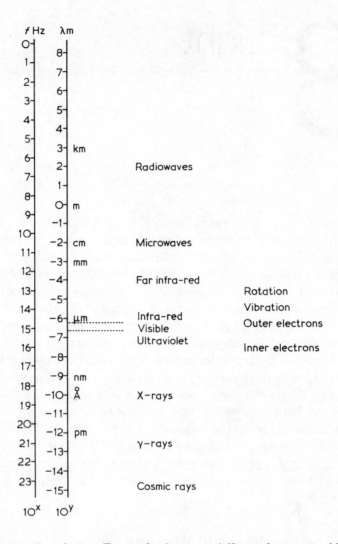

Fig. 13.1 Electromagnetic radiation. Types of radiation at different frequencies (f) and wavelengths (λ). The numbers on the scales indicate the power of ten. Thus the frequency of visible light is about 10^{15} Hz with a wavelength of about 500 nm. The right hand column indicates the type of molecular energy levels involved for absorption or emission of light at the wavelengths indicated.

Another property of light is that its energy is quantized (see Chapter 14). This means that the intensity of a beam of light cannot be simply reduced in a continuous fashion until it is zero, but rather that there is a minimum energy, known as a *quantum*. The unit of light, with one quantum of energy, is known as a *photon*. A light beam is made up of a number of photons. Of course, in normal bright light the number of photons involved is so great that it is impossible to distinguish a change in intensity produced by one photon; however the human eye is capable of detecting

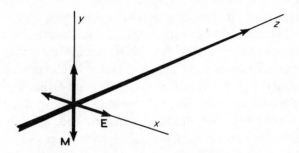

Fig. 13.2 A light beam travelling in the z direction has an electric vector oscillating in the x-direction, and a magnetic vector oscillating in the y-direction.

light sources giving off only a few quanta per second, and under these circumstances, such differences are detectable. At this level the release of individual photons becomes a statistical event, as does the absorption of a photon by the visual pigment of the eye.

The energy of a single quantum of light is related to its frequency by

$$\text{Energy} = h\nu \qquad (13.2)$$

where h is a constant, known as Planck's constant after the physicist who first suggested this relationship. Under no circumstances is it possible to subdivide a quantum of light. For example, light absorption occurs because a particular atom or molecule of the absorbing material is excited into a high energy state, by a single photon. As discussed in Chapter 14, for this to happen the energy of the incoming light quantum must exactly equal the energy change necessary to excite the molecule.

The intensity of a beam of light is measured in terms of the energy per second crossing unit area of cross section perpendicular to the direction of the beam. The units of light intensity are thus watt/metre^2.

You can get some idea of the different energies associated with the different quanta at different wavelengths by considering the ability of electromagnetic radiation to blacken photographic film. This is blackened by all radiation of wavelengths less than a certain threshold (about 1 μm) however low the intensity, but by no wavelengths above this threshold, however great the intensity. This is because there is insufficient energy in each quantum of the long-wavelength to excite the chemical process underlying the photographic process (the ionization of the silver halide in the film).

13.1   Colour

Different colours of visible light are simply light of different wavelengths. The human eye is sensitive to wavelengths of electromagnetic ratiation between about 400 nm (blue) and 750 nm (red). If all wavelengths within this range are present at equal intensity then the light is said to be 'white'. Objects illuminated with white light in general appear coloured. The reason for this is that certain wavelengths are absorbed by the object (as discussed in Chapter 14) due to the presence of suitable energy levels of excitation of the electrons of the object. Other wavelengths are reflected, and it is the wavelengths that are reflected that give the object its colour. Thus chlorophyll absorbs both red and blue light, and thus appears as (white minus red minus blue) which appears green. An object which looks red is one which absorbs all wavelengths less than about 660 nm — and thus reflects all light at longer wavelengths.

Ultraviolet is the name given to light of shorter wave than that visible to the human eye, and infra-red to that light of longer wavelength than is visible to the eye.

The eye is able to detect colour because the light sensitive cells of the eye involved with colour vision (the cones) contain one of three different pigments. The pigments differ in that they absorb light of different wavelengths, and thus different cones are excited by different wavelengths of light. The sensation of colour can arise because the connections of the cones to the brain are wired in a highly specific manner.

Three terms are used to describe colours; intensity, hue and saturation. *Intensity* is the energy of the light falling on a detector in unit time. The *hue* describes the wavelengths involved. The *saturation* is a measure of the degree to which the colour is 'mixed' with white light. Thus pure red is a saturated colour, whereas pink is an unsaturated red, being a mixture of red and white light.

When a surface is illuminated by light of a particular range of wavelengths, then some light will be absorbed, depending upon the pigments in the surface, and some light reflected. If no light is reflected the surface appears black. The range of wavelengths that are reflected obviously depends upon the range of wavelengths in the illuminating light. For this reason objects can appear to be coloured differently when viewed in light of different colours — for example an object which looks green in white light will look black when viewed in red or blue light (since that object absorbs both red and blue light.)

If two coloured lights are shone onto a white surface, then the colour seen will be that sensed when the relevant cones in the eye are illuminated by the sum of the two lights (colour addition). Thus green and red lights mixed appear yellow (because yellow light excites green and red cones of the eye equally, so a mixture of red and green light causes the same sensation).

If two coloured pigments are mixed, then the resulting colour will be that of the light which is still not absorbed (colour subtraction). Thus if a green pigment and a red pigment are mixed, blue and red light will be absorbed by the green pigment, and all except the red light will be absorbed by the red pigment. The precise colour seen will depend upon the degree of absorbance of the two pigments over the entire spectrum. Because this will not be perfect the resulting colour will not be black, but generally appears a dark brown.

13.2 Polarized light

We have seen that light can be considered in terms of an electric disturbance which is oscillating sinusoidally in a plane perpendicular to the direction of travel of the light. This refers to a single quantum. In a light beam containing many millions of quanta per second, the plane of polarization of each can be random, and thus there is no preferred direction of orientation. When this is the situation the light is said to be *unpolarized*. This is the normal situation in everyday life, see Fig. 13.3.

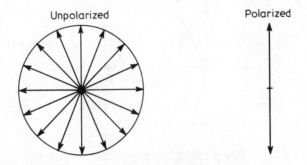

Fig. 13.3 Unpolarized and polarized light. The figure illustrates the directions of the electric vector of a ray of light propagated into the page.

In order to understand the various concepts of polarized light it is important to appreciate that the electric disturbance can be considered as a vector oscillating in a plane perpendicular to the direction of travel of the light, and that like other vectors discussed in Chapter 2, can be resolved into components. For example, the vector E_r in Fig. 13.4 represents the electric vector of a light beam which is directed into the paper, at an instant of time. This vector can be resolved into two components at right angles to one another, as indicated by E_1 and E_2. From the behaviour of the light beam it is impossible to distinguish between these descriptions of the light.

It is possible to remove the components of a beam of light that are not polarized in a particular direction, and thus to produce *polarized* light. This can be done with a variety of mineral crystals and reflection from surfaces produces partially polarized light. Nowadays polarized light is almost always produced in the laboratory by a material known as 'Polaroid' which is made up of long hydrocarbon chains containing iodine, oriented in one direction and embedded in a sheet of gelatin. The direction of the components of light which are transmitted by the Polaroid is known as the *axis* of the sheet of Polaroid. Thus, for example, if the light polarized in the direction of E_r of Fig. 13.4 is directed at a piece of polaroid whose axis is oriented parallel to E_1, then the light transmitted through the polaroid is the component E_1, (of amplitude $E_r \cos\theta$); the component E_2 (of amplitude $E_r \sin\theta$) is absorbed. The plane of polarization has thus been changed (and the intensity reduced – see Fig. 13.5). The

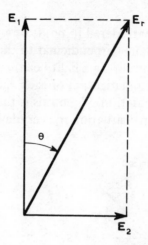

Fig. 13.4 The vector $\mathbf{E_r}$, representing the electric vector of a light beam at an instant of time, can be resolved into the two vectors $\mathbf{E_1}$ and $\mathbf{E_2}$.

intensity of the transmitted beam is the square of the amplitude. Thus:

$$\text{intensity transmitted} = \text{incident intensity} \times \cos^2\theta \qquad (13.3)$$

This is known as *Malus' law*.

13.3   Dichroism and birefringence

Dichroism and birefringence are two related phenomena. They concern anisotropy of materials – i.e. different properties of the same materials in different directions. *Dichroic* substances are those which transmit polarized light incident in one plane of polarization, but absorb polarized light whose plane of polarization is in a different direction. Dichroic substances can obviously be used to produce polarized light, and polaroid, discussed above is dichroic. *Birefringent* substances are those which have a refractive index which is different for different planes of polarization. In one direction the refractive index will be maximal, (often perpendicular to the first), and in another direction it will be minimal. These two directions are known as the *principal axes*, and it is easiest, when considering the effects of birefringent substances, to think in terms of the components of the light polarized in the directions of the principle axes. Having different refractive indices in different directions means that the velocity of the light is different in these different directions but the frequency is unchanged, and thus that the wavelength is different.

We shall see (page 204) that the refractive index of a material varies with the wavelength of the light used to measure the refractive index. With birefringent materials both refractive indices are wavelength dependent. It also happens that dichroism and birefringence are not as separable as they seem on the above definitions, since a material which is birefringent at one wavelength may be dichroic at another.

Birefringence and/or dichroism occur whenever a number of long-chain molecules are aligned parallel to one another. Biologically the importance of these phenomena is that the occurence of such alignment can be detected. The alignment can occur naturally, or it can be induced under specific experimental conditions.

An example of naturally occuring birefringence is provided by striated muscle. When viewed under a microscope with a polarizer in the condenser and an analyzer in the eyepiece, striated muscle shows a strong banding pattern. The regions where the birefringence is very strong are known as A bands (A for Anisotropic), and the regions with low birefringence are known as I (for Isotropic) bands.

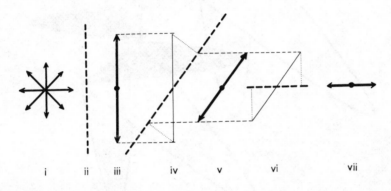

i   ii   iii   iv   v   vi   vii

Fig. 13.5 (i) Unpolarized light is passed through a sheet of polaroid whose axis is in the direction indicated by (ii). The result is light with the resultant vector (iii). Note that the amplitude of the light is twice as great as the amplitude of the light in any direction before polarization. If the polarized light (iii) is passed through a sheet of polaroid whose axis is in the direction (iv), only the component (v) in this direction is propagated. The result is a rotation of the direction of polarization, together with a reduction in intensity. The procedure can be repeated (vi) and (vii). Note that the direction of polarization in (vii) is perpendicular to that in (iii).

An example of a situation in which birefringence can be induced is that of a solution containing molecules which are highly asymmetrical in shape − that is, they are much longer in one direction than in others, like a rod rather than a sphere. Under normal conditions the molecules will be aligned randomly in the solution, which therefore shows no birefringence. However, if the solution is made to flow down a tube, the effect of the flow will be to cause an alignment of the molecules in the direction of the flow. The degree of alignment will be greater the greater the velocity. Furthermore, the alignment will be greater, at a given velocity, the greater the degree of asymmetry in the molecules. By measuring the birefringence of the solution as it flows down the tube it is therefore possible to estimate the shape of the molecule. The way in which the degree of birefringence is estimated is to polarize the light

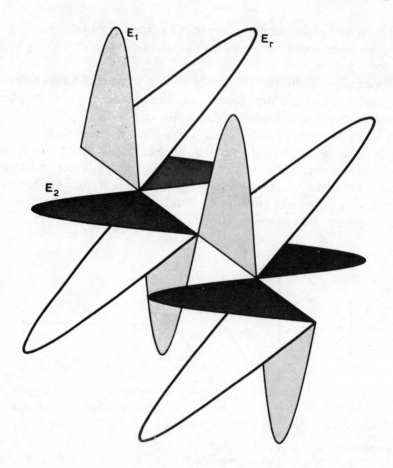

Fig. 13.6A Plane and circularly polarized light. Two beams of light travelling along the same path, and of equal intensity, have electric vectors E_1 and E_2, polarized in mutually perpendicular directions. When the two vectors are in phase then the resultant E_r is plane polarized light, polarized at an angle of 45° to that of E_1 and E_2.

falling on the solution with a piece of polaroid (the polarizer) whose axis is 45° to the direction of flow, and to view the flow through a second piece of polaroid (the analyzer) whose axis is perpendicular to the axis of the polarizer. The background will appear dark. Birefringent materials will appear light (the reasoning is discussed in Fig. 13.5), the intensity being greater the greater the degree of birefringence of the solution. This method of estimating the shape of molecules is known as *flow birefringence.*

13.4   Plane and circularly polarized light. Wave plates

We saw above that we could consider polarized light in terms of a vector, and that we

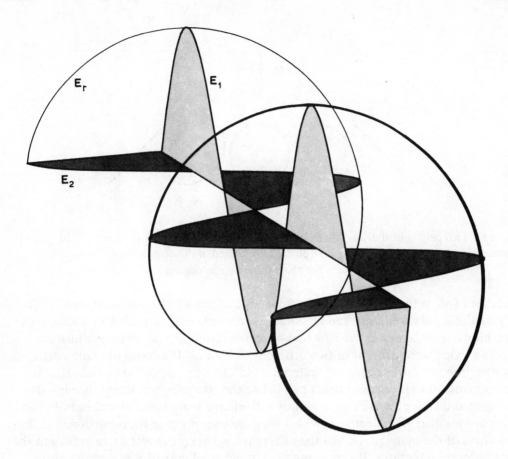

Fig. 13.6 B. The two beams are one quarter of a wavelength out of phase, then the resultant vector
E_r is a helix, and the resultant is known as circularly polarized light.

could treat this vector in the same way as other vectors, and find the component of
the vector in different directions. Suppose, as in Fig. 13.6A that we have two beams
of polarized light whose vectors are perpendicular to one another. If the two vectors
are in phase (i.e. the peak of the sinusoid of one vector occurs at the same instant
of time as the peak of the sinusoid of the other vector) then the resultant is a plane
polarized beam at 45° (if the amplitudes of the two beams were equal) or at some
other angle if the amplitudes were different. However, suppose that the two beams are
90° out-of-phase, so that the peak of one of the sinusoids occurs at the zero of the
other, as in Fig. 13.6B then the resultant will be a vector which rotates in a circle if
the two amplitudes were equal, or in an ellipse if they were unequal. This resultant is
known as *circularly polarized light*.

If the amplitudes of the two beams we are considering is the same, then as the
phase varies, so the resultant beam changes from being plane polarized (when they
are in phase) through regions of being elliptically polarized to being circularly

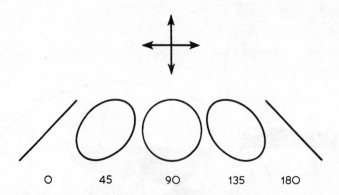

Fig. 13.7 Two perpendicularly polarized beams of light (top row) have a resultant whose type of polarization depends upon their relative phase, as indicated in the bottom row. The numbers indicate the phase difference in degrees.

polarized (when the beams are exactly 90° out of phase). This is illustrated in Fig. 13.7.

Circularly polarized light can be obtained from plane polarized light with the use of a birefringent material. We saw that a birefringent material was one whose refractive index was different in two different directions. If a beam of plane polarized light is shone at the birefringent material at 45° to its principal axes, then the two components of that incident beam parallel to the two axes will travel through the material at different velocities, and thus will emerge from the material out-of-phase with one another. The degree of which they are out of phase is proportional to the thickness of the material, and to the difference between the refractive indices in the two different directions. By choosing the correct thickness of material the phase difference can be made of any desired value. It is possible to buy plates, known as *wave plates* whose difference in optical thickness in the two directions is a specified proportion of the wavelength of light of some particular wavelength. Quarter and half-wave plates are common. A quarter wave-plate is one for which beams of polarized light parallel to the principal axes which were originally in-phase, emerge one-quarter of a wavelength (or 90°) out of phase. Thus a quarter wave plate will produce circularly polarized light from plane polarized light incident at 45° to its principal axes.

The difference in the optical thickness in the two directions of a birefringent material will be different for different wavelengths. Thus the type of polarization emerging from the material if white plane-polarized light is incident on the material will be different for different colours. If the light is then passed through a piece of polaroid some colours will be absorbed and some transmitted. The result will thus be to produce coloured light. An effective demonstration of this is to take one or more layers of sellotape, and place this between crossed polaroids (polaroid sunglasses will be satisfactory) with the axis of the sellotape at 45° to the polaroid.

Different colours will be seen according to the number of layers. Sellotape is made from a cellulose tape which is composed of long hydrocarbon chains oriented along the length of the sellotape, and which is therefore birefringent.

13.5 Interference of light

In many situations light will arrive at a point in space from several directions. Provided that:

1. The beams are of light of the same frequency.
2. The beams are coherent (this means that the relative phases of the beams is constant, which in effect required that the beams originate from the same source),

then the observed amplitude of light at that point is the sum of the amplitudes of the light from the several different directions. We have seen above that we can consider light in terms of an oscillating electric vector, whose instantaneous amplitude is given thus by $A \sin \omega t$. If we consider two beams of light of amplitudes A_1 and A_2, with a phase difference of ϕ, then the resultant is

$$A_1 \sin \omega t + A_2 \sin(\omega t + \phi) \qquad (13.4)$$

If the phase difference ϕ between the two beams is zero, then the resultant is a beam of amplitude equal to the sum of the amplitudes of the two separate beams, which are then said to interfere constructively. If the phase difference ϕ is 180°, so that the beams are exactly out of phase, then the resultant has an amplitude equal to the difference between the amplitudes of the two beams and the interference is said to be destructive. In this case if the amplitudes are equal, then the resulting intensity at that point will be zero.

13.6 Diffraction

An understanding of diffraction of light (and other forms of electromagnetic radiation) is of importance in at least three regions of biology. The limit to the resolution of the microscope is set by the effects of diffraction of light at the lens; spectroscopy requires monochromatic light of varying wavelength, and diffraction gratings are one means of producing this; X-ray crystallography, which has enabled the structures of many biologically important macromolecules to be determined, is a problem of diffraction of X-rays by crystalline arrays of those molecules.

If a parallel beam of light is directed at a fairly wide slit and then viewed on a screen, an image of the slit will be formed which is about the same width as the slit. If the slit is now made narrower and narrower, the image will not continue to

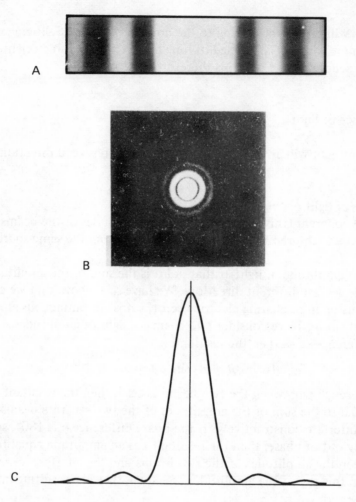

Fig. 13.8 A. Diffraction pattern obtained when a parallel beam of light is passed through a narrow slit. B. Similarly formed diffraction pattern but with a small hole as the apperture. This is known as an Airy Disc. C. Distribution of intensity across such diffraction patterns.

remain the same width as the slit, but instead a wider *diffraction pattern* will be obtained locking like that shown in Fig. 13.8A.

In order to understand the production of such diffraction patterns (which seem to violate the idea that light travels in straight lines) it is easiest to consider light from yet another standpoint. This is that a beam of light can be thought of as a series of wavefronts. Every point in a wavefront is to be considered a secondary source of light, producing a spherical series of wavefronts propagating from that point, as indicated in Fig. 13.9. If there are no obstacles in the way, then the result of having a whole line of such secondary sources will be to produce a plane wavefront in front of the one being considered, since the components in all other directions will interfere

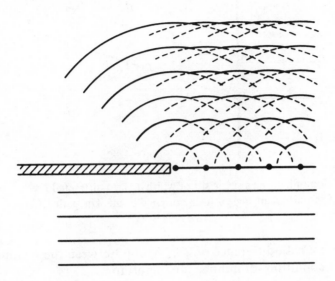

Fig. 13.9 Propagation of wavefronts. The lines are drawn one wavelength apart and represent the wavefronts of a parallel beam of light. Any point in the wavefront can be considered as a secondary source of light. A series of such points is illustrated. The resultant of those in the body of the beam is still a parallel beam, since the components in other directions interfere destructively. However, at an edge this does not happen and the light spreads into the region behind the object producing the edge.

destructively with one another. However, at the edge of such a wavefront, the components on one side, necessary for this destructive interference, will be absent, with the result that light will be propagated into the region that would otherwise be dark.

The reason for the particular diffraction pattern of Fig. 13.8 can now be understood by reference to Fig. 13.10. Provided that the screen is a long way from the slit then the rays of light originating from different points within the slit that hit the same point on the screen can be considered to be parallel. Consider the two extreme rays originating from either side of the slit, and making an angle θ with the direction of the original beam. The ray from the lower edge will have to travel a distance YZ further than the top ray before it hits the screen. This distance YZ = $a \sin \theta$. If YZ = λ, the wavelength of the light used, then the ray from the top of the screen X and the central ray have a path difference of $\lambda/2$ and hence annul one another at the screen. The ray immediately below X will be annuled by the ray immediately below the central ray and so on — for every ray in the top half of the slit there will be a ray in the lower half to interfere destructively with it at the screen. Thus, when $a \sin \theta = \lambda$ there will be zero intensity on the screen. The same will hold

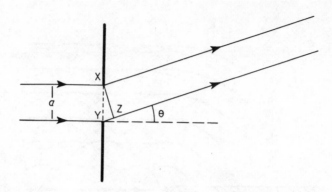

Fig. 13.10 Diffraction of light be a single slit of width a. Two diffracted rays are shown, originating from either side of the slit, at an angle θ to the original beam. The path difference between these two rays is YZ.

true of course for a path difference of 2λ, 3λ, etc. Between these minima there will be maxima. The condition for minima are therefore

Minimum $\qquad a \sin\theta = n\lambda \qquad n$ a positive integer $\qquad$ (13.5)

or $\qquad\qquad\qquad\qquad \sin\theta = \dfrac{n\lambda}{a} \qquad\qquad\qquad\qquad$ (13.6)

For a circular hole of diameter D instead of a slit the derivation is more complex, but a diffraction pattern as illustrated in Fig. 13.8B is formed, whose minima occur at

Minima when: $\qquad\qquad\qquad \sin\theta = \dfrac{1 \cdot 22 n\lambda}{D} \qquad\qquad\qquad$ (13.7)

Thus for small holes, the smaller the hole the larger the diffraction pattern.

The diffraction pattern from a circular aperture is known as an *Airy disc.*

13.7 Refractive index

Light travels at different velocities through different media. Its velocity is greatest in a vacuum. The ratio of its velocity *in vacuo* to that in the medium is known as the refractive index of the medium. The magnitude of the refractive index is dependent upon the wavelength of the light being used. The refractive index of air is very close to 1.

As the light passes through the medium, its frequency does not change. From equation 13.1 it follows therefore that the wavelength of the light must decrease as the velocity decreases.

For dilute solutions the refractive index of the solution is proportional to the concentration of the solute. Thus by measuring the refractive index an estimate of concentrations can be made. Use is made of this relationship in the ultracentrifuge, in experiments in which concentration gradients are obtained along the length of the centrifuge tube.

13.8   Optical thickness

If a substance of refractive index μ has a thickness t then in the time that it takes a ray of light to cross from one side of the substance to the other a ray of light in vacuum would have travelled μt. This is known as the optical thickness of the substance.

13.9   Refraction and reflection

1. A ray of light reflected from a surface lies in the plane containing the incident ray and the normal to the surface. The angle of reflection equals the angle of incidence. See Fig. 13.11A.

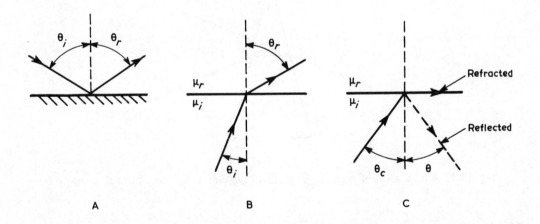

A   B   C

Fig. 13.11 A. Reflection of light from a surface. B. Refraction of light at a boundary between two media of refractive index μ_i and μ_r. C. The critical angle θ_c is obtained when the refracted ray leaves the higher refractive index medium at $90°$. The dashed line indicates the reflected light.

2. A ray of light refracted at the boundary between materials of different refractive index lies in the plane containing the incident ray and the normal to the surface. If

θ_i is the angle of incidence and θ_r is the angle of refraction, then:

$$\mu_i \sin \theta_i = \mu_r \sin \theta_r \qquad (13.8)$$

where μ_i is the refractive index of the material of the incident ray, and μ_r is that for the refracted ray. This is known as *Snell's law*. See Fig. 13.11B. If μ_r is less than μ_i, then a *critical angle* is reached when θ_r equals 90°

$$\sin(\theta_{\text{i. critical}}) = \mu_r/\mu_i \qquad (13.9)$$

For angles greater than the critical angle then *total internal reflection* occurs. (See Fig. 13.11C). Use is made of this fact when using a prism as a reflecting surface, as for example in binoculars or a periscope.

13.10   Thin prism

Provided the angle of the prism A is less than about 10°, and also that the angle of incidence of a ray of light onto the surface of the prism is small then (see Fig. 13.12).

$$D = (\mu - 1)A \qquad (13.10)$$

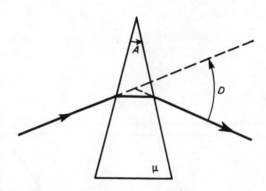

Fig. 13.12 A thin prism of angle A. An incident ray will be deviated through an angle D.

13.11   Dispersion

The deviation produced by a thin prism is dependent upon the refractive index of the material used to make the prism. Since this refractive index will be different for different wavelengths (colours), it follows that the deviation will be different for different colours. It is for this reason that a prism can be used, for example, to form a spectrum of the sun's rays, and why prisms can be used in spectrophotometers.

The *dispersive power* of a substance is a measure of its ability to separate different wavelengths, and is defined as

$$\text{Dispersive power} = \frac{\mu_b - \mu_r}{\mu_y - 1} \tag{13.11}$$

where the subscripts b, r and y refer to blue, red and yellow light (of particular wavelengths).

13.12   Thin lens

A lens is a piece of transparent material with spherical surfaces; various shapes of lens can be made — see Fig. 13.13. A thin lens is defined as one whose thickness is

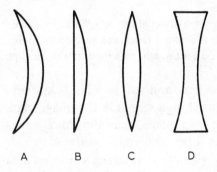

Fig. 13.13 Different shapes of lens. A. Concave-convex. B. Plano-convex. C. Biconvex. D. Biconcave.

much smaller than the radii of curvature of its faces. We shall only consider lens in air. Lens have the property of focusing and image formation. The *focal length* of a lens is defined as the distance between the centre of the lens and that point at which rays parallel to the axis of the lens are brought to a focus, and the *focal point* is that point (see Fig. 13.14). There is a focal point either side of the lens. The focal length is a function of the refractive index of the material of the lens (μ) and of its radii of curvature

$$\frac{1}{f} = (\mu - 1)\left(\frac{1}{r_1} + \frac{1}{r_2}\right) \tag{13.12}$$

in which r_1 and r_2 are the two radii of curvature, defined as positive for a convex surface and negative for a concave surface. It is possible to build lens of the same focal length but with different radii of curvature.

A lens will form an image of an object placed one side of it. We shall only consider objects placed on or near the axis. Provided that the object is placed further away from the lens than the focal plane (Fig. 13.15A) then the image will be formed

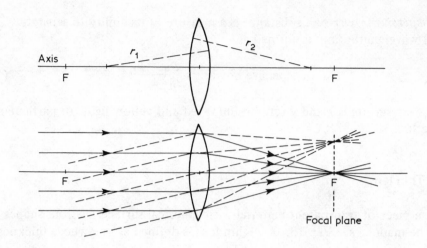

Fig. 13.14 A. A lens is made from a material having spherical surface with radii of curvature r_1 and r_2. B. Light parallel to the axis is brought to a focus at the focal point. Parallel light, at an angle to the axis, is brought to a focus in the focal plane.

on the opposite side of the lens, and will be what is known as a 'real' image, meaning that the image can be focused on a screen in the image plane. If the object is placed closer to the lens than the focal plane, then the lens is acting as a magnifying glass, as shown in Fig. 13.15B. The image is formed on the same side of the lens as the object and is what is known as a 'virtual' or 'imaginary' image, meaning that the image cannot be visualized on a screen, but that the rays of light leaving the lens can be extrapolated backwards as indicated, and behave as if they were originating from the virtual image. If a second lens (such as the lens of the eye) is placed to the right of the magnifying glass, it can produce a real image of the virtual image, formed by the magnifying glass.

If u is the distance between the object and the centre of the lens, called the object distance, and v is the distance between the image and the centre of the lens, called the image distance, then

$$\frac{1}{u} + \frac{1}{v} = \frac{1}{f}$$

(13.13)

If the image is a real image, then v is positive; if the image is a virtual image then v is negative.

Ray diagrams for a lens can readily be drawn, and the method is illustrated in Fig. 13.15A, B. The diagram is drawn for rays originating from the top of the arrow. The ray which passes through the centre of the lens (2) passes through undeviated. The ray which runs parallel to the lens axis must pass through the focal point behind the lens (1), and the ray which passes through the focal point in front of the lens must be refracted so as to then run parallel to the axis (3). These three rays meet at a

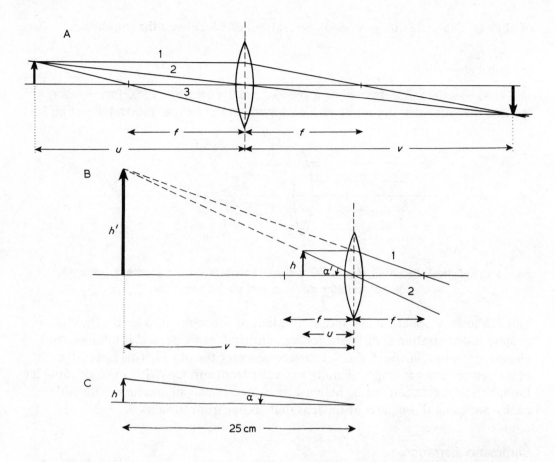

Fig. 13.15 Image formation by lens. A. Use of lens to form a real image. The object is placed further from the lens than the focal point. B. Use of lens as a magnifying glass, producing a virtual image. The object is placed between the focal point and the lens. Note how the rays drawn from object follow the rules for drawing ray diagrams. The dashed lines are extrapolated. C. The same object as in B, placed 25 cm from the viewing point, subtends an angle α.

point, thereby producing the image in the image plane.

If two thin lens are placed in contact with one another, then the focal length of the combination (f) is given by:

$$\frac{1}{f} = \frac{1}{f_1} + \frac{1}{f_2} \tag{13.14}$$

where f_1 and f_2 are the focal lengths of the separate lens.

13.13 Lens aberrations

Aberrations in lens are features which prevent a lens from producing perfect images of objects. There are many types of aberration, of which we shall consider only three.

Spherical aberration

In a lens, rays that are refracted near the axis are brought to a focal further away from the lens than rays that are refracted far away from the axis, as illustrated in Fig. 13.16.

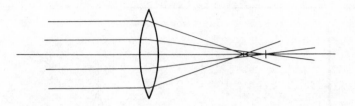

Fig. 13.16 Spherical aberration. Rays close to the axis are brought to a focus at a point more distant from the lens than rays far from the axis.

This is known as spherical aberration. For lens of the same focal length the amount of spherical aberration is different for lens of different shape, and is minimal for plano-convex lens in which the flat surface is nearer the object. This is way the object lens of a microscope generally has a flat front surface. With any individual lens, the spherical aberration can be reduced by using as small an aperture as possible — i.e. by not using those parts of the lens that are far from the axis.

Chromatic aberration   ·

Because the refractive index of any material is different for different wavelengths of light it follows that the focal length of a lens (from equation 13.12) will vary with the different wavelengths. Therefore, using a simple lens, an object viewed in white light will have its outline blurred and coloured. This is known as chromatic aberration. Chromatic aberration can be corrected for two wavelengths by using a doublet. This is a combination of two lens in series, one of which is diverging and the other is converging. The two lens must be made of glass with different refractive indices — crown glass and flint glass are often used. The dispersion produced by the one lens can then be corrected by the other.

Astigmatism

This is one of several lens defects that occur for objects that are viewed a long way from the lens axis. The effect is an inability to focus points. Astigmatism is a fairly common defect in the human eye. Its presence here is detectable by looking at a set of lines radiating out from a common point, as in the spokes of a bicycle wheel.

To the astigmatic eye only those in a certain direction can be focused properly; those in other directions appear blurred. It can be corrected by wearing glasses containing a cylindrical lens over the astigmatic eye.

13.14   Least distance of distinct vision

The human eye can focus objects at different distances by a process known as accommodation, in which the shape of the lens is changed so as to change the focal length. The normal eye can accommodate readily down to objects placed about 25 cm away from the eye, and this is known as the least distance of distinct vision.

13.15   Magnification

Fig. 13.15B illustrates the use of a simple lens as a magnifying glass. Depending upon the distance between the lens and the object the virtual image can be made bigger or smaller. At the same time the image will be nearer to or further away from the lens, and thus to the eye which is placed immediately behind the lens. There is a variety of expressions to describe the relative size of the object and image. The *lateral magnification* is the ration of the height of the image to the height of the object.

$$\text{Lateral magnification} = \frac{h'}{h} \qquad (13.15)$$

The *magnifying power* is the ratio of the angle subtended by the eye of the image, when the image is formed at the least distance of distinct vision (25 cm), to the angle subtended at the eye by the object when the object is placed at the least distance of distinct vision.

For the simple lens used as a magnifying glass the magnifying power can be readily determined. If the image is to be formed 25 cm from the lens, then the object distance (in cm) is given by

$$\frac{1}{u} = \frac{1}{f} + \frac{1}{25} \qquad (13.16)$$

The angle subtended at the eye by the object is then given by $\tan\alpha = h/u$. For small values of α, $\tan\alpha \doteq \alpha$, and thus the angle subtended at the eye by the image, when the image is situated at the least distance of distinct vision is

$$\alpha = \frac{h}{u} = \frac{h}{f} + \frac{h}{25} \qquad (13.17)$$

The angle subtended by the object without the lens when the object is at the least distance of distinct vision is (Fig. 13.15C)

$$\alpha' = \frac{1}{25} \tag{13.18}$$

and thus the magnifying power is

$$\text{M.P.} = \frac{\alpha}{\alpha'} = \frac{25}{f} + 1 \tag{13.19}$$

If the image is located at infinity (which means that the object is placed at the focal point of the lens), then the angular magnification in this case is given by:

$$\text{angular magnification (image at infinity)} = \frac{25}{f} \tag{13.20}$$

With high powered lens for which f is much less than 25 cm, this angular magnification is approximately the same as the magnifying power and the magnification indicated on, say, a microscope eyepiece, is obtained from the formula with the object at infinity. It is more relaxing to use a magnifying glass or a microscope set up with the object at infinity than with the object at the least distance of distinct vision. A 10X eyepiece is common, for which the focal length therefore is $25/10 = 2\cdot5$ cm.

13.16   The compound microscope

In the compound microscope a short focal length lens (the objective) is used to form a real image of an object, and this is then magnified by an eyepiece lens used as a magnifying glass. The two lens are held a fixed distance apart, and focusing is achieved by changing the distance between the specimen and the objective lens. Fig. 13.17 gives a ray diagram for the microscope. In this instrument the objective lens is in reality a compound lens with several components, designed to reduce aberrations to a minimum. The eyepiece is also complex.

For high quality microscopy the form of illumination of the object is crucial. Two forms of illumination are *critical* and *Köhler* illumination. In critical illumination (Fig. 13.18A) a diffuse source of light is focused with a condenser lens onto the object plane (the plane containing the object). Before the advent of electric power this was the type of illumination used; necessarily so in order to provide sufficient intensity. It was thought at the time to be critical to obtain good focusing of the light source onto the object plane for optimal resolution. This is now known not to be the case.

In Köhler illumination (Fig. 13.18B) the light source is focused by a lens onto the back focal plane of the condenser lens. The light from any point in the back focal plane emerges from the condenser lens as a parallel beam, and is focused again in the back focal plane of the objective lens, diverging again to be diffuse in the image plane of the objective lens. Köhler illumination is the form of illumination used nowadays in research microscopes.

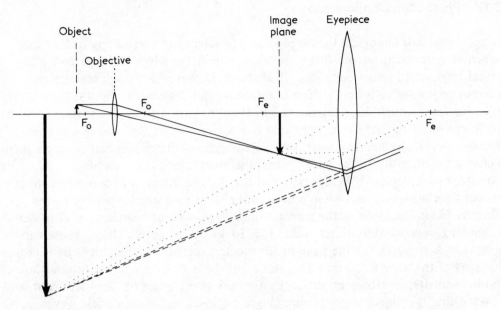

Fig. 13.17 The compound microscope. The solid lines are two rays from the object followed through to the eyepiece. The dashed lines are extrapolations of the solid lines to the final image. The dotted lines demonstrate the image formation by the eyepiece using the image plane as the object, and can be compared with the rays of Fig. 13.15B.

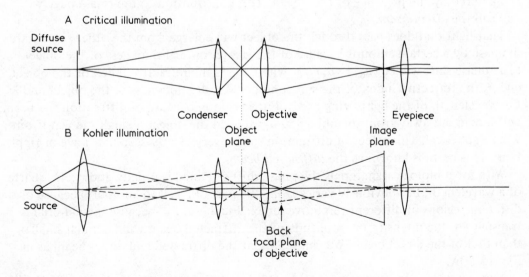

Fig. 13.18 A. Critical illumination. B. Köhler illumination. The solid lines represent light from the source which are not scattered or diffracted by the object. In critical illumination the diffuse source will be in focus with the object. The dashed lines in B represent light scattered by the object, plus extrapolations of these rays back through the condenser lens.

13.17   Phase-contrast microscopy

Living, unstained, biological tissues present a problem for microscopy in that they are generally transparent and therefore show little detail when investigated by the normal bright-field microscopy described above. However, although transparent, different regions of cells have different refractive indices, and the light is slowed down whilst passing through these regions. Light that has passed through the tissue thus has a different phase to that of the light that has not passed through the tissue. Phase contrast microscopy makes use of these relative phase differences and converts them to changes in intensity in the image plane, thereby making them visible.

   In order to understand the operation of the phase-contrast microscope we must first consider what happens when a beam of light meets a small region of higher refractive index than that of the surroundings. The incident beam can be considered as a moving wave-front as illustrated in Fig. 13.19 by the parallel lines. These can be thought of as representing the peak of the electric vector of the incident light beam. That part of the wave front that does not meet the region of higher refractive index continues unaltered. However, diffraction occurs at the edges of the wave front, and provided that the object is not too much greater in diameter than a few wavelengths of light (not larger than say a few μm) then the diffraction of that part of the incident wavefront that does not pass through the region of higher refractive index causes the wavefront to 'join up' again in front of the object, as indicated in Fig. 13.19. Since none of this light has passed through the region of higher refractive index, its phase is unaltered by the presence of the object. This contribution to the final image we will call the *direct beam*.

   That light that does pass through the object will emerge from the other side retarded in phase by a certain amount because of the greater optical thickness of the object. (The phase shift will be $(\mu_{0b} - \mu_s) \, t/\lambda$, where μ_{0b} is the refractive index of the object, and μ_s the refractive index of the surroundings, t is the thickness of the object and λ the wavelength of the light being used). Furthermore, because of diffraction the beam will spread out away from the object, and provided the object is small the wavefront will be spherical. This effect of diffraction can be very clearly seen in a wave or ripple tank. This beam is known as the *diffracted beam*.

   With most biological materials the phase shift obtained is usually about one quarter of a wavelength, but obviously varies from one part of a cell to another, because of the different regions of different refractive index. Furthermore, because the material is transparent, the intensity or amplitude of the diffracted beam will be much smaller than that of the direct beam. We can represent the diffracted and direct beam as in Fig. 13.20A.

   Now suppose that in some way we could retard the diffracted beam by a further one quarter of a wavelength, it would then be directly out-of-phase with the direct beam, (Fig. 13.20B), and would interfere destructively in the image plane. In order to produce darkness in the regions of high refractive index it is necessary also to reduce

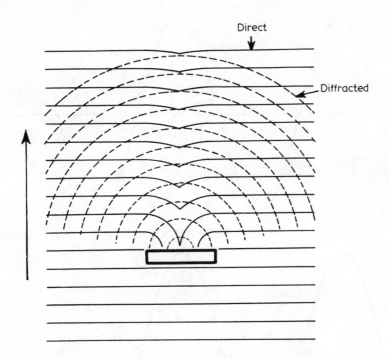

Fig. 13.19 The solid lines represent the wavefronts of light which has not passed through the object, travelling in the direction of the arrow. The lines are drawn one wavelength apart. Several wavelengths in front of the object this direct beam has parallel wavefronts once again. The dashed, circular lines represent the wavefronts of light diffracted by the object. Note that these are retarded behind the direct beam.

the intensity of the direct beam (Fig. 13.20C) to be approximately equal to that of the diffracted beam. If this is done, then an image is formed in the image plane whose intensity varies as the refractive index of the sample varies. For optical thickness differences of exactly one quarter of a wavelength, or of one and a quarter wavelengths etc. then the intensity will be a minimum. For differences in optical thickness less than or greater than one quarter of a wavelength the intensity will be greater, and will be greatest for phase differences through the object of three quarters of a wavelength. Note that good phase contrast will only be obtained for objects whose regions of equal refractive index are fairly small. For larger objects the phase contrast will only appear at the edges of the material. For this reason artifacts can readily be seen in cells with large inclusions – e.g. extra bands can be seen under phase contrast in muscle cells with long sarcomeres.

With the use of Köhler illumination it is possible to separate the direct and diffracted beams spatially in the back focal plane of the objective, and thus to change their relative phase and to reduce the intensity of the direct beam. Suppose that the back focal plane of the condenser contains a mask with a small hole at its centre, as

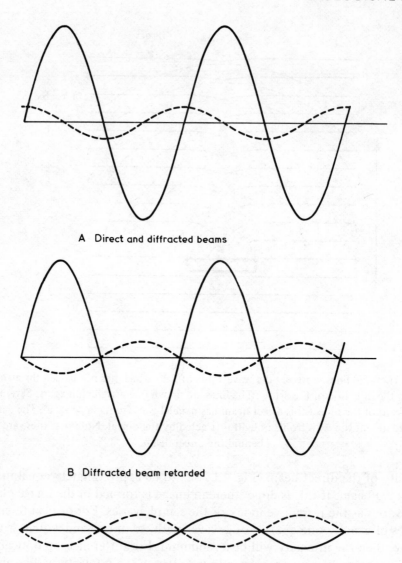

A Direct and diffracted beams

B Diffracted beam retarded

C Intensity of direct beam reduced

Fig. 13.20 A. The direct and diffracted beams from an object such as that illustrated in Fig. 13.17. The diffracted beam (dashed line) is of lower intensity, and retarded (in this case by one quarter of a wavelength) behind the direct beam (solid line). B. The effect of retarding the diffracted beam by a further quarter of a wavelength. The beams are now exactly out-of-phase. C. The effect of reducing the intensity of the direct beam.

illustrated in Fig. 13.21. The light passing through the object will be parallel (since it originates from the focal point of the condenser lens), and in the absence of an object will therefore pass through the image of the hole at the focal point in the back

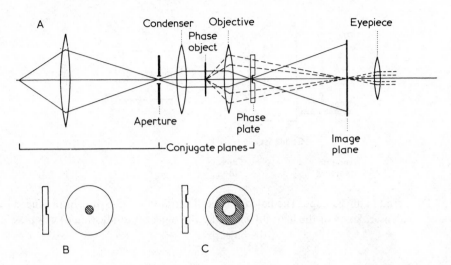

Fig. 13.21 Phase contrast microscopy. A. The diagram is basically that of Fig. 13.16B with an aperture in the focal plane of the condenser lens, and a phase plate in the back focal plane of the objective. B. The shape of the phase plate used for the description in the text, and in the upper part of the figure. The hatched area is thinner than the rest of the plate and covered by a material to reduce the intensity of light passing through this region of the plate. C. The shape of the phase plate (and aperture) normally used. The operation is the same as that for the phase plate illustrated in B.

focal plane of the objective, as shown. However the diffracted light, originating at the object, will in general not pass through the image of the hole, but will be spread over a wide area in the back focal plane of the objective. Thus, in the back focal plane of the objective we have good spatial separation of the two beams. In this plane we therefore insert a piece of material which retards the light by one quarter of a wavelength everywhere except at the image of the hole; in this region we position a substance which will reduce the intensity of the light without causing any phase change (a thin coating of metal is often used). The insertion is usually known as a *phase plate*. We have now effected the operations necessary for phase-contrast microscopy.

In fact, a small hole as the light source in the back focal plane of the condenser results in very low intensities of the final image. Instead it is normal to use an annulus as the source of light in the condenser (Fig. 13.21C); the phase plate to be inserted into the back focal plane of the objective must now contain an annulus of intensity reducing substance with a phase retarder everywhere else, this annulus covering the image of the source annulus in the condenser.

13.18   Dark-ground and interference microscopy

Two other forms of microscopy which enable transparent objects to be seen are

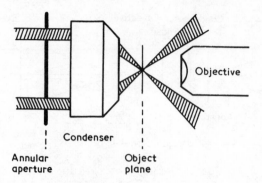

Fig. 13.22 Dark-ground microscope. The hatched areas illustrate the cone of light produced by a dark-ground condenser. Some of the light diffracted by the object (not illustrated) will pass into the objective.

dark-ground and interference microscopy. We discussed in the previous section how the light proceeding from such an object could be treated in terms of a direct and a diffracted beam. In dark ground microscopy the condenser is designed (see Fig. 13.22) so that the direct beam misses the objective altogether, and the only light entering the objective is the diffracted light. The condenser is designed therefore to produce a cone of light. The background thus appears black, and the objects which under normal bright-field microscopy would appear transparent, show up clearly.

Interference microscopy is beyond the scope of this book. The principle is to split the light from the condenser into two paths (with the use of a half-silvered mirror), one of which passes through the object and the other of which does not. These two paths are recombined before the objective, and thus interfere with one another. The phase of the light taking the path away from the object can be caused to change its phase relative to the other path by a precisely known amount, and by this means accurate estimates of the optical thickness of the specimen can be determined.

13.19   Polarization microscopy

The object of such microscopy is to show up regions of dichroism or birefringence. The principle is straightforward. A polarizer is introduced on the condenser side of the object (usually below the condenser lens), and an analyzer is inserted after the object, often above the objective. The relative alignment of the analyzer and polarizer is variable, but they are usually used with their axes at right angles. A wave-plate is also often inserted between the analyzer and polarizer, in order to produce a coloured image, as described on page 200.

Although the principle is straightforward a good polarizing microscope needs all its lens to be strain free (otherwise the strain lines distort the image), and these are expensive. Furthermore the object stage needs to rotate precisely about the axis of the objective, in order that the object can be rotated relative to the plane of polarization of the incident light without moving out of the field of view.

13.20   The resolving power of the light microscope

In the treatment above of the diffraction caused by a single hole we depicted the situation in which a parallel beam of light was incident upon a small hole. The diffraction pattern was caused by the interference of the light originating from different regions of the hole and which travelled different distances to points on the pattern. Although the geometry of the microscope is different (we have to consider a point source placed close to a comparatively large lens) the image of the point source is a similar diffraction pattern known as an Airy Disc. The cause of this diffraction pattern, as before, is the different distances that individual rays travel between leaving the point source and arriving at a particular point in the image plane, and their resulting interference. With the treatment of the light diffracted by a small hole, we saw that the size of the diffraction pattern was inversely proportional to the size of the hole. In the case of the microscope the aperture of the lens is the equivalent of the hole, and we find that the size of the Airy Disc is inversely proportional to the angle subtended at the lens by the object (the angle α in Fig. 13.23). For a lens of a given focal length, the larger the lens, the smaller the diffraction pattern.

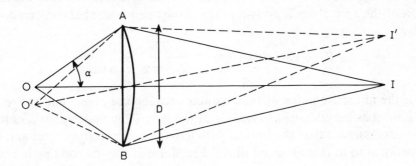

Fig. 13.23 The resolving power of the light microscope. Two point sources at O and O' will from Airy discs centred at I and I'.

This diffraction pattern sets a limit to the resolving power of the microscope. Let us determine how close two point sources of light can be so as to be resolvable. In Fig. 13.23 these are O and O'. Each point source will produce its own diffraction pattern, centred about the points I and I'. If these diffraction patterns overlap there will come a separation at which it is no longer possible to tell that there are two discs rather than just one. This will be a subjective measure. The *Rayleigh criterion*

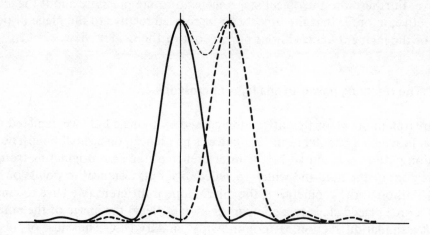

Fig. 13.24 Rayleigh's criterion for two Airy discs to be resolvable. The maximum of the one falls over the first minimum of the other. The summed intensity in the central region is illustrated also. For narrower separations the resulting pattern is said to be not resolvable.

says that the two Airy Discs will be just resolvable if the maximum of the one falls over the first minimum of the other (Fig. 13.24). To be just resolvable therefore, the angular separation of the two point sources of light is equal to the angular separation of the first minimum of the Airy Disc from the centre of the pattern.

The spatial separation (s) of the two point sources giving rise to this angular separation of the Airy Discs is known as the resolving power of the microscope, and is given by:

$$\text{Resolving power } (s) \;=\; \frac{1\cdot22\lambda}{2\mu\sin\alpha} \;=\; \frac{0\cdot61\lambda}{\mu\sin\alpha} \tag{13.21}$$

where μ is the refractive index of the medium between the object and the lens. $\mu\sin\alpha$ is known as the *numerical aperture* of the lens. (The reason that μ enters into the above expression is that the optical path lengths between the object and the lens are proportional to μ. It is assumed that the medium in the microscope is always air with a refractive index of 1.)

Sinα cannot be greater than 1 (in which case $\alpha = 90°$), and thus the numerical aperture is limited by the refractive index of the medium. Oil is often used (with what is known as an oil immersion objective) to increase the numerical aperture, and thus to provide an improved resolving power. A good high-powered oil immersion objective will have a numerical aperture of about 1·3. The resolving power of the microscope is thus about one half of the wavelength of the light used. Greater resolution is thus obtained by using light of short wavelengths (i.e. blue rather than red light). The electron microscope provides such an improvement in resolution by virtue of the fact that it uses electrons of extremely short wavelength.

13.21   The diffraction grating

A diffraction grating is a large number of slits, regularly spaced, with a spacing not much greater than the wavelength of the light. See Fig. 13.25.

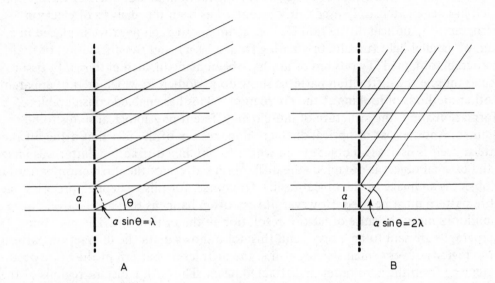

Fig. 13.25 The diffraction grating. Light incident on the grating will be diffracted through all angles as it leaves the grating. Rays diffracted through an angle θ only are illustrated.

Consider a parallel beam of light striking the grating. The path difference between rays diffracted at an angle θ to the grating is $a \sin \theta$, where a is the spacing between the lines of the grating. If the light is viewed a long way from the grating compared with the spacing, then these rays will all meet in the same place (to be strictly accurate the rays meeting in the same place will all have angles very close to θ). The rays will interfere to produce a maximum if the path difference is a whole number (n) of wavelengths:

$$\text{Maxima when} \qquad a \sin \theta = n\lambda \qquad (13.22)$$

Provided the number of lines in the grating is very large, then these maxima will be sharply defined lines. Note that the smaller the spacing a, the larger the angle θ required to produce a given order of diffraction for light of a particular wavelength.

The light which is not diffracted is known as the zero order diffraction. The maximum that occurs with $n = 1$ is known as the first order; that with $n = 2$ the second order diffraction, etc.

The above discussion has considered the diffraction pattern caused by an object with a single periodicity. If the object has more than one periodicity, then each one can give rise to its own diffraction pattern. In this case note that the largest spacings on the diffraction pattern correspond to the smallest spacings in the original object.

The importance of diffraction studies lies in the fact that from the diffraction pattern it is in principle possible to determine the structure of the material giving rise to that diffraction pattern. Extensive use of this fact has been made in the study of the structure of macromolecules by X-ray diffraction. Another use of diffraction which has been extensively used fairly recently has been the analysis of electron micrographs by optical diffraction. The electron micrograph negative is placed in a beam of parallel light (usually originating from a laser), and the diffraction of that micrograph obtained. Two types of use have been made of such patterns. In tissues such as muscle the diffraction patterns show up periodicities which are not apparent to other methods of looking at the micrographs. This has enabled considerable information about the structure of muscle in different physiological states to be obtained. Another use of optical diffraction makes use of the fact that if the diffraction pattern itself is used as an object from which to obtain a diffraction pattern, as image of the original object is obtained (the diffraction pattern of the diffraction pattern of an object is an image of the object itself). Of course, for this to happen precisely the entire pattern must be used. However, it very often happens that an electron micro-graph looks messy because of lack of resolution in the instrument. The resulting micrographs are said to be 'noisy', and this noise shows up in the diffraction pattern in the regions of very small spacings – i.e. the diffraction pattern of the noise occurs a long way from the zero order undiffracted beam. Thus if the distant regions of the original diffraction pattern are not used to recreate the image of the object, the noise is removed, and it is often possible to obtain information about regularly occuring structures in the original object which are not immediately apparent in the original micrograph.

13.22   Interference colours from thin films. Structural colours

A thin film is a sheet of material whose thickness is comparable to that of the wave-length of light, i.e. of the order of 1 μm. Araldite sections for electron-microscopy, films of oil on water, soap bubbles, scales on the wings of certain species of butter-fly, elytra of some beetles, and the materials giving rise to the greens of most birds are examples of structures having colours produced by thin films or related methods of colour production.

Fig. 13.16 illustrates the way in which such interference colours are produced. Light is incident on a thin film of different refractive index to its surroundings. Some of the light will be reflected at each boundary. The result will be a set of beams of reflected light which will thus be in a condition to interfere with one another if brought to a focus by, say, the eye.

The optical path difference between adjacent rays 2 $\mu t \cos(r)$, can be determined in the following way. In Fig. 13.26B the line EK is drawn perpendicular to the two

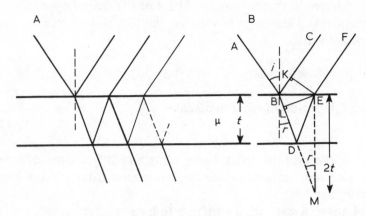

Fig. 13.26 Reflections by thin films. The left hand side illustrates the way the reflections arise. The right hand side illustrates the geometry as described in the text.

emerging beams, and the difference in optical paths of these two beams is (BD + DE) − BK, where BK is in air, and the other beam is in the material of refractive index μ. If we draw EL perpendicular to the line BD, then it is a property of refracted rays that in the time it takes for the refracted ray to travel the distance BL, the reflected ray has travelled the distance BK. The optical path difference is thus (LD + DE). In Fig. 13.16B, EM is drawn perpendicular to the surface of the film. DM = DE. Thus LM = (LD + DE) = $2t\cos(r)$.

A further property of reflection at surfaces is that there is a phase change of π (or half a wavelength) for reflections at a boundary between a region of low refractive index and one of high refractive index, but not for reflections at a boundary between a medium of high refractive index and one of low refractive index. Thus when the optical path difference between the two rays is an integral number of wavelengths, the interference will be destructive rather than constructive. Thus we obtain:

$$\text{Minima when } 2\mu t\cos(r) = n\lambda \qquad (13.23)$$

For any particular angle of incidence some wavelengths will interfere constructively and some destructively, resulting in coloured effects with white light incident upon the film. The colours will change with the angle of incidence, which is why one sees complete spectra in such things as soap bubbles and oil-on-water. Such colours (which change as the angle of viewing is changed) are said to be *iridescent*.

We have described above how colours are formed with single thin films. Very often in animals there are several layers of thin films of equal spacing superimposed on one another. The effect is to augment the intensities of the reflected colours; the above treatment is still valid, although a difference in the spacing of the regions of low and high refractive index will complicate the colours seen.

Note that if the optical path difference is a single wavelength of light then the rate of change of colour with angle of incidence will be much less than if the optical path difference is a larger number of wavelengths. For the closest approach to a single colour, independent of the angle of viewing, the film should thus be about half a wavelength of light thick.

13.23 X-ray diffraction of macromolecules

We saw in the section on the diffraction grating how diffraction patterns could be formed from electron micrographs, or for that matter from other objects containing regular periodicities.

In fact it is not necessary for a structure to have regular periodicities for it to be able to form a diffraction pattern. Atoms will diffract X-rays, and since X-rays have wavelengths of the order of a few tenths of a nanometre, which is comparable to interatomic spacings, diffraction patterns can be obtained of molecules. The intensity obtainable from a single molecule is much too low to be detectable, and therefore X-ray diffraction patterns are obtained from crystals of molecules. The limiting factor at the present day to the study of biologically interesting molecules is the ability to crystallise those molecules.

Two types of information is present in such a diffraction pattern; information about the structure of the molecule itself and information about the structure of the crystal (the way in which the molecules are packed to form the crystal). In order to understand how these two types of information are separable, Fig. 13.27 illustrates some optical diffraction patterns of 'molecules' which are just spots on paper. A single 'molecule' gives rise to the type of diffraction pattern shown in Fig. 13.27A. This diffraction pattern is what is necessary to determine the structure of the molecule. Fig. 13.27B illustrates the diffraction pattern formed from a crystalline array of single points – the diffraction pattern is also a series of points, whose separation is related to the separation of the points in the object by equation 13.22. If we now look at the diffraction pattern of the array of molecules in Fig. 13.27C we see that this has the same position of points of that formed from the crystal lattice of the single points, but that the intensity of the diffraction pattern is that produced by the single molecule. In other words the diffraction pattern of the single molecule has been *sampled* at a series of positions determined by the structure of the crystal lattice. The X-ray crystallographer has to infer the diffraction pattern of the single molecule from the sampled pattern that he can obtain. Thus, of the two kinds of information available in the pattern, the crystal lattice of structure is obtained from the position of the spots, and the molecular structure from the intensities of those spots.

A further problem for the crystallographer is that the only methods for detecting the spots at his disposal are ones which measure the *intensity* of the radiation at that point. In order to work out the structure the crystallographer needs to know the *amplitude* of the radiation. The difference arises from the fact that the amplitude at one point can be positive whilst at another point it is negative. The intensity, being the square of the amplitude, does not indicate the sign of the amplitude, and this information is therefore lost. The normal way round this problem is to take diffraction pictures of two forms of the crystal — the normal form, and a form into which a heavy atom has been added at some suitable point in the molecular structure. The heavy atom with its greater ability to diffract X-rays causes sufficient change in the diffraction pattern, with relatively little change in the molecular structure, to enable the sign of the amplitude to be determined by comparing the two patterns.

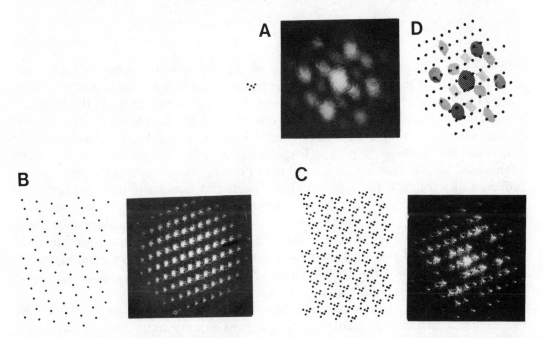

Fig. 13.27 A. The pattern with the dark background is the diffraction pattern of the 'molecule' simulated by the points shown to its left. B. A lattice of points gives rise to a diffraction pattern which is also a lattice of points (the blurred edges arise because the lattice is small — a larger lattice would give rise to sharper points). C. A lattice of the molecules of (A) has points in the same positions as for the point lattice (B), but with intensity of points governed by the intensity of the molecular diffraction pattern (A). D. illustrates the superposition of the lattice points on the molecular diffraction pattern.

14 The Interaction of Light and Matter

The way that light interacts with matter is of importance to biologists for a number of reasons. At the top of the list must be because the ability of plants to absorb the energy of the light from the sun, and convert it into chemical energy in the form of organic molecules, is fundamental to the maintenance of life. There are other examples of light absorption and conversion to chemical energy of which the process of vision is one, and the synthesis of vitamin D in the human skin another. Furthermore, both animals and plants make a great deal of use of colour, mainly connected with the processes of warning, camouflage and identification; a number of different techniques are used for producing colours. Analytically the biochemist is also very concerned with the interaction of light and matter, in that different molecules absorb light in highly characteristic ways, and this enables both quantitative and qualitative measurements to be made of the contents of solutions. The general field of such study is known as spectroscopy. Finally, a number of animals have evolved mechanisms which convert chemical energy into light.

An understanding of the way that light (and other wavelengths of electromagnetic radiation) interact with molecules requires at least a rudimentary understanding of the nature of atoms and molecules. Much more complete accounts of this subject will be found in textbooks of physical chemistry, and indeed much of the subject matter of modern physics (meaning 20th century physics) is concerned with this problem.

14.1 Electronic energy levels

The particular property of atoms and molecules that we need to understand is that of *energy levels.* An atom can be thought to consist, for the purposes of this chapter, as a nucleus with a positive charge surrounded by a number of electrons, each with a negative charge. Despite the strong electrostatic attraction arising from the opposite charges of the nucleus and the electrons, the electrons are constrained to orbits around the nucleus. Within the orbits the attractive electrostatic forces between the nucleus and the electron are balanced by the centrifugal forces of the motion of the electron. The full description of the orbits and the energies of the electrons is the subject of quantum or wave mechanics.

A fundamental postulate of quantum mechanics, first expounded by Planck, and for which there is no explanation, is that the energy of electromagnetic radiation (this includes light) is quantized — meaning that the energy of, say, a beam of light is made up of units known as *photons* and that the minimum amount of light that can be obtained with that wavelength is one quantum. The energy of the quantum is related to the wavelength of the light:

$$E = h\nu = Hc/\lambda \qquad (14.1)$$

in which ν is the frequency, λ the wavelength and c the velocity of the light, and h is Planck's constant ($6 \cdot 63 \times 10^{-34}$ Js).

The effect of this postulate (together with another that an orbiting electron radiates electromagnetic radiation when it jumps from one orbit to another, at a frequency related to the frequencies of rotation in the two orbits) is that the electrons can only have discrete energies or, as it is usually expressed, can only exist in a number of discrete energy levels. Furthermore, at any instant of time no more than two electrons can occupy the same energy level (and if two electrons are occupying the same energy level then they have opposite values of *spin* — rotating about their own axis either clockwise or anticlockwise).

For any given atom there is a number of possible energy levels, those with the lowest energy being the ones in which the electron is closest to the nucleus. Because the amount of energy that must be given to an electron to cause it to jump to a higher level is much greater than thermal energy (kT) at room temperature, it follows that electrons usually occupy the lowest available energy levels. Fig. 14.1 illustrates the energy levels of a hypothetical atom, and their occupancy by seven electrons. The two electrons in each of the three lowest levels have been depicted by arrows pointing in opposite directions to indicate their opposite spins. There is in fact an infinite number of possible energy levels, whose separation becomes less and less. They are indicated by the hatched area in the figure. Energies greater than a certain level — the *ionization level* — cause the electron to be no longer attached to that particular atom, and the atom is then said to be ionized.

If by some means an electron is given sufficient energy (for example light, heat, electrical or chemical energy) then it becomes able to jump from one energy level to a higher one absorbing the energy in the process. For this to occur it must be provided with precisely the correct amount of energy — if the amount is too large then the transition will not occur. (If the amount of energy is sufficient to enable the ionization level to be reached, then any amount of energy greater than this value will cause ionization, the excess energy being converted to kinetic energy shared between the electron and the atom.)

Likewise, when an electron jumps from a high energy level to a lower one, then the energy difference of the electron in the two states is released. This will often be in the form of light energy, in which case one quantum of light will be emitted. The

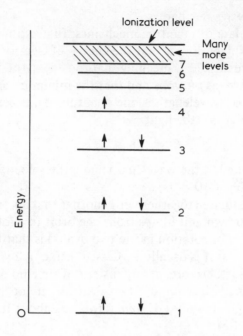

Fig. 14.1 The energy levels of a hypothetical atom.

wavelength of the light emitted under these circumstances is dependent upon the energy difference between the initial and the final states of the electrons, and is given by:

$$\lambda = \frac{hc}{\Delta E} \qquad (14.2)$$

in which ΔE is the difference in energy between the two states. When an electron in a low energy state is excited into a higher energy state, with the absorption of one quantum of light, then the wavelength of the light necessary to allow this transition is also given by the above formula. Equations 14.1 and 14.2 are one and the same relationship, bearing in mind that the frequency and wavelength of light are related by

$$\lambda = c/\nu \qquad (14.3)$$

Thus, as we might expect, the transitions between energy levels of an electron require the absorption, or give rise to the emission, of one quantum of light of that wavelength whose energy equals the difference in energy between the two states.

Another way that an electron can lose its energy, other than by the emission of light, is by transferring the energy to an adjacent molecule which has two energy levels separated by the same energy difference. This molecule can be of the same or a different species. This method of energy transfer is of the utmost importance in biological systems; thus the ability of chlorophyll to make use of the energy absorbed

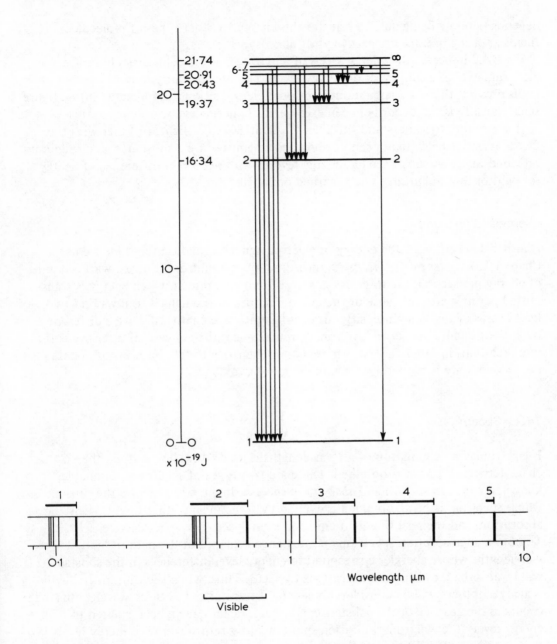

Fig. 14.2 A. The energy levels of the hydrogen atom, drawn to scale. Only the first seven levels are indicated. The vertical, arrowed lines represent possible electronic transitions between these seven states. The vertical line on the right hand side indicated the ionization energy from state 1. B. The spectrum of hydrogen. The spectral lines form series (labelled 1–5 in the diagram), corresponding to transitions from the state given by the number to higher numbered states. Thus in the group of lines labelled '2' the transitions are from state 2 to states 3, 4, . . . etc.

depends upon its being able to transfer this energy to other types of molecule, thereby setting in train the pathways of photosynthesis.

Fig. 14.2 indicates the energy levels of a hydrogen atom. This atom has only one electron, and has a simple set of energy levels. Despite the fairly limited number of levels present, there is a large number of possible transitions, as indicated in the figure, since transitions can occur between every pair of energy levels.

It is essential to understand this idea of quantization of light, together with the fact that only single quanta can be absorbed or emitted by a substance, and then only provided that that substance has energy levels with the correct difference of energy for light of that quantum to be absorbed or emitted.

Energy and intensity

Another description of the energy of electromagnetic radiation (and one we use in Chapter 13) is that of the *intensity*, measured in watt metre^{-2}. Since a watt is a unit of energy per second, the intensity is a measure of the rate at which energy is transmitted per unit area. Because of the quantal nature of the light it follows that two light beams of the same intensity but of different wavelength will have a different number of quanta per second. In some circumstances the concept of intensity is the more relevant, in other cases it is more comprehensible to talk in terms of quanta per second. Care is needed not to muddle these two concepts.

14.2   Spectra

If electromagnetic radiation of all wavelengths is used to excite an atom, then a characteristic *spectrum* is obtained. There are two types of spectrum obtainable, known as *emission spectra* and *absorption spectra*. In emission spectra the substance whose spectrum is being obtained is excited by some means other than by light absorption, and the resulting light emission is analysed by a spectroscope.
To obtain an absorption spectrum the substance is placed in the path of light. Those wavelengths whose energies correspond to energy level differences in the substance will be absorbed, causing the transitions to occur. Thus, when the transmitted light is analyzed, there will be considerably less light transmitted at these wavelengths than occurs in the absence of the substance. Of course, after having been excited to higher energy levels, the electrons will often spontaneously return to lower energy levels, thereby giving off light, and at first sight it might appear that this emitted light will be of the same intensity as the absorbed light, and thus cause no absorption to be detectable. The significant point here is that the light which is subsequently emitted is emitted in all directions, whereas the original beam of light was highly directional. Thus the intensity, as detected in one direction only will be much lower. The spectrum of hydrogen is depicted in Fig. 14.2B.

The types of spectra that are obtained in this manner, whether absorption or emission spectra, are known as *line spectra*, since the spectra occur as a number of discrete lines. These are to be distinguished from *continuous spectra*, which are obtained, for example, from a normal tungsten filament electric light bulb, and are caused by black-body radiation as discussed in Chapter 9. Continuous spectra contain all wavelengths within a certain range, as indicated, for example, in Fig. 9.3.

The spectrum of the hydrogen atom is as simple a spectrum as can be obtained. Only one electron is involved and thus there is a limited number of well-spaced lines. When we consider the situation that occurs in more complex situations then the spectra likewise become more complex. There are several sources of complexity: 1. Atoms with more than one electron. 2. Molecules rather than atoms. 3. Spectra of atoms or molecules in the liquid or solid states rather than the gaseous state that we have been assuming (without having said so explicitly) up until now.

14.3   Rotational and vibrational energy levels

We shall not consider the case of atoms with more than one electron. However, when we come to consider molecules we immediately introduce two more sources of energy levels; these are concerned with the interactions of the atoms of the molecule, and are *vibrational* and *rotational* energy levels. If we consider the case of a diatomic molecule (see Fig. 14.3) then the atoms are separated by a distance determined by the balance of attractive and repulsive forces. The atoms can vibrate backwards and forwards relative to one another (as though they were connected by a spring) and can also rotate about a common axis. There will be a certain amount of energy involved in both the vibration and the rotation, and the total energy of the molecule will include the energy from these two modes as well as from the electronic energy levels.

For reasons that are beyond the scope of this book to explain (but which can be found in the textbooks on physical chemistry cited in the Bibliography at the end of the book) only certain discrete energy levels are available to the molecule for either rotation or vibration. The situation is exactly analogous to the electronic energy levels discussed above. We therefore talk about the vibrational energy levels, and the rotational energy levels of the molecule. At any instant of time a particular molecule must possess a certain energy of vibration, and thus be in one of its vibrational energy levels, and likewise be in one of its rotational energy levels. As with any type of energy level, the distribution of molecules with different values of energies will be given by the Boltzman distribution.

The magnitude of the energies associated with the transitions between rotational and vibrational energy levels are less than those associated with the transitions of electrons between their energy levels. The spectral lines associated with such transitions will thus be at longer wavelengths and lower frequencies, i.e. in the infra-red

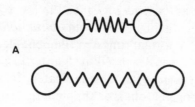

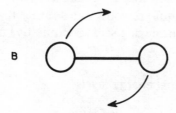

Fig. 14.3 A. Vibration in a diatomic molecule. Two separations of the atoms in a hypothetical vibration state are shown. The relative force between the molecules varies as the separation changes. B. Rotation in a diatomic molecule.

region of the spectrum (see Fig. 13.1). Vibrational energy level differences are greater than those of transitions between rotational energy levels.

At any instant of time a molecule which has a particular electron in a particular energy level may be in any of its rotational or vibrational energy levels. If we consider the total energy of a molecule, then we can think of this as being made up of the sum of the energies associated with the various classes of energy level. Just considering the rotational, vibrational and electronic energy levels (we shall see below that there are other forms), then

$$E_{\text{total}} = E_{\text{rotation}} + E_{\text{vibration}} + E_{\text{electronic}} \qquad (14.4)$$

The full energy level diagram of a molecule including these three types of energy will thus look something like that shown in Fig. 14.4.

The precise rules by which it can be determined whether or not any particular transition can occur between any pair of energy levels are beyond the scope of this book. However, in general, transitions can occur between any pair of energy levels. Thus, when we come to consider transitions between two electron-energy-levels, there will in fact be a variety of possible energy changes, due to the presence of the vibrational and rotational energy levels superimposed on any electronic energy level. This will result in a series of fairly closely spectral lines. Organic molecules containing large numbers of atoms will thus have very complex spectra, associated with a large number of electronic transitions, each of which can be associated with a large number

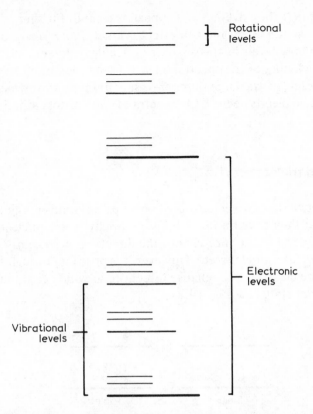

Fig. 14.4 The energy diagram of a molecule indicating (not to scale) the range of energy levels when electronic, vibrational and rotational energies are taken into account. As drawn the molecule has available two electronic states, three vibrational and three rotational states.

of possible rotational and vibrational levels.

14.4   Molecules in solution

A further influence upon the spectra associated with a particular substance is the state that substance is in — whether it is present as a gas or in solution or in a solid. The spectral lines of gases tend to be very sharp, the molecules being far enough apart to exert no influence upon one another. However, in the liquid phase the molecules are close together. There are several consequences of this. In the first place, the rotation of the molecules is no longer unhindered, and the effect of this is to modify the rotational energy levels of any particular molecule very slightly. Since this modification is random, the appearance of the spectrum (which is the summed response of a very large number of molecules) is that the spectral lines which were discrete with the molecules in the gaseous state are now broadened out to such an extent that they overlap and appear as a band. Further, if the molecules are sufficiently close together (and this effect is much more prevalent in the solid state), then the energy levels of the separate molecules interact, and the result is also seen as a broadening of the spectral lines. For complex molecules in solution the result is a merging of the large number of spectral lines into wide bands; for example, this can be clearly seen in the spectra of the photosynthetic and visual pigments.

14.5   Singlet and triplet states

We mentioned above that two electrons can occupy any one energy level at any one time provided that they have opposite spins. Normally when an electron is excited to a higher energy level it does not change the direction of its spin — the excited state is then known as a *singlet* state. However it is possible, though far less probable, that the spin of the electron does change direction, in which case the excited state is known as a *triplet* state (see Fig. 14.5).

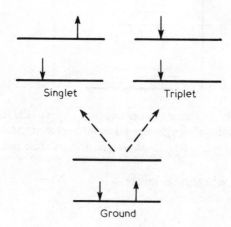

Fig. 14.5 Excitation of a ground state to a singlet excited state (on left) and triplet excited state (on right). The direction of the arrow indicated the direction of spin. In a triplet transition both the state and the spin must change.

After a transition of an electron to an excited state, there is a high probability that the electron will return to its unexcited or *ground state*, emitting radiation. The excited state is described as having a *lifetime*, which is the average time that an electron will spend in the excited state before it decays to the ground state. The lifetime of singlet states is of the order of $10^{-9} - 10^{-8}$ s. Triplet states, having a much lower probability of making the transition (due to the fact that in order to return to the ground state the excited electron must also reverse the direction of its spin since two electrons in one state must have opposite spins), have very much longer lifetimes, typically of about 10^{-5} s, although very much longer lifetimes, up to seconds long are known.

TABLE 14.1

Atomic and molecular times (seconds)	
Orbital period of electron around nucleus	10^{-16}
Vibrational period	10^{-13}
Lifetime of vibrational excitation	10^{-12}
Molecular collisions every	10^{-11}
Lifetime of electronic excitation	
— Singlet	$10^{-9} - 10^{-8}$
— Triplet	10^{-5} — several seconds

Note that these times are only rough approximations, and represent average times.

The times involved in molecular events are of vital importance in understanding the types of reactions that occur. Table 14.1 lists some typical times. For example, it is possible for the energy of a molecule which has been sent into an excited electron energy level state by the absorption of light to lose that energy to another molecule, rather than by the emission of electromagnetic radiation, and to thereby excite that second molecule. For this to happen to any appreciable extent, the time involved in transferring the excitation to the second molecule must be much faster than the lifetime of the excited state. The longer the lifetime of the excited state, then the higher the probability of transferring the energy in some form other than by re-radiating the energy as electromagnetic radiation. It is for this reason that the idea that a triplet state is involved in photosynthesis is so attractive.

14.6 Fluorescence and phosphorescence

When an electron is caused to jump from its ground state to a higher energy level, then, generally, the vibrational state will change to a higher energy level also. This is shown in Fig. 14.6. If the excited state is a singlet state, then the average lifetime of the state is about 10^{-8} s. In solution the lifetime of an excited vibrational energy level is much less than this, and thus the vibrational energy will be lost before the electron returns to its ground state. When the electron does eventually return to its ground state, then the energy released will be slightly less than that necessary to cause the initial excitation, since some of that energy has already been lost. The wavelength of the emitted photon will thus be slightly longer than that of the exciting photon (a shift towards the red end of the spectrum). This is known as

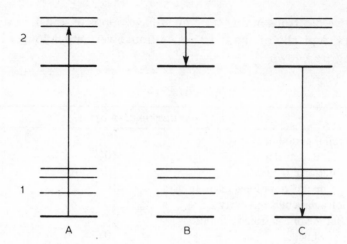

Fig. 14.6 Fluorescence. A. Excitation, from the ground state of both the electronic and vibrational levels to excited states of both vibrational and electronic levels. B. Decay of excitation from the excited vibration state to the ground vibration state. C. Subsequent decay from the excited electronic state to the ground electronic state.

fluorescence. If the excited state of the electron is a triplet state, then the time before the return to the ground state occurs is much longer, and the phenomenon is then known as *phosphorescence*.

Because the emitted photons are at a longer wavelength than the exciting photons it is straightforward to detect them. Furthermore, since light can be detected at very low intensities using photomultipliers, experiments which make use of the phenomenon of fluorescence are extremely sensitive. The use of fluorescent probes, as well as of the intrinsic fluorescence of certain of the amino-acids of proteins (e.g. tryptophan) has become an important biochemical tool in recent years. There have been two general fields of study:

1. If the orientation of the molecule excited by a photon does not change before the fluorescence occurs, then the orientation of the polarization of the emitted photon will be the same as that of the exciting photon. Thus, by using polarized light to excite the molecule it becomes possible to detect movement in the molecule during the time required to fluoresce. With singlet states this means that the rotation of molecules in periods of about $10^{-8} - 10^{-9}$ s can be detected. This technique is known as *fluorescence depolarization*.
2. If the environment of the fluorescent molecule changes, then changes in the form of the fluoresence also occur. These may be shifts in the wavelength of the emitted photon, or changes in the width of the spectrum of the emitted light. By this means it is possible to detect chemical changes. This has been particularly useful in the study of the properties of membranes, into which it has become relatively straightforward to position fluorescent probes.

14.7   X-rays

In atoms with many electrons the electrons generally occupy the lowest available
energy levels. When discussing the excitation of an electron it is generally implied
that we are talking about the electrons in the highest energy levels — the valence
electrons. However, it is possible to excite electrons from lower energy levels above
the valence electrons, or, more usually, to ionize the atom by providing an inner
electron with sufficient energy to leave the atom altogether (see Fig. 14.7). If this
happens then the lower energy level is quickly filled by an electron from a higher
energy level, usually the electron in the highest energy level. Since the energy level
difference between the states will be large, the energy of the photon emitted will
likewise be large, resulting in a photon with a very short wavelength, usually of the
order of a few tenths of a nanometre. Such photons are known as *X-rays*. The normal
method of producing X-rays is to bombard a metal (usually copper) with very high
energy electrons (produced from a heated filament and accelerated down a large
voltage gradient). Provided that these have sufficient energy they will be able to
knock the inner electrons from the copper, thereby allowing the production of
X-rays.

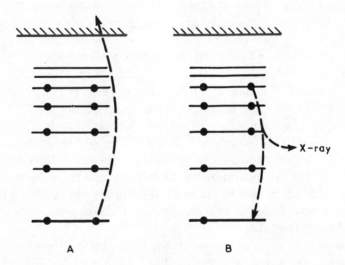

Fig. 14.7 Production of X-rays. A. An electron from a low energy level is knocked out of the atom.
B. An electron in a higher energy level decays to the low level with the production of an X-ray.

14.8   Induced radiation and lasers

An electron in an excited state will normally decay spontaneously to the ground
state, releasing a photon of light at the frequency determined by equation 14.1. However the decay can be induced by radiation of the same frequency. An important

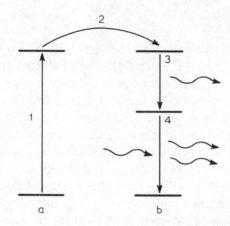

Fig. 14.8 Laser action. 1. Substance a is excited to a high energy state. 2. The excitation is transferred directly to substance b. 3. This decays to an intermediate energy level spontaneously, releasing a quantum of light. 4. This intermediate state is induced to decay by incoming light.

property of the induced light is that its phase and direction are the same as that of
the inducing light. Thus, if by some means a substance can be excited to a particular
energy level which then normally decays to its ground state releasing a particular
wavelength of photons, then shining a beam of light of that wavelength into the substance will cause the induced decay of the excited molecules, thereby augmenting
the beam. Lasers (which stand for *L*ight *A*mplification by *S*timulated *E*mission of
*R*adiation) make use of this phenomenon, see Fig. 14.8. They contain a substance
which can be excited, by some means, to a particular energy level. The high-energy
state then decays, releasing photons of a particular wavelength. The substance is
held in a tube with mirrors at each end. Some of the light from the decay is thus
reflected from the mirrors and as it traverses the tube causes the stimulated emission
of radiation. Provided the separation of the mirrors, and their alignment, is accurate
the result is to build up an intense, parallel, light beam. One of the mirrors is half-
silvered so that a part of the beam is transmitted through the mirror rather than
reflected back along the tube.

The method of excitation of the substance of the laser differs from one type to
another. For example, in the helium-neon laser often used for diffraction experiments in biological laboratories, the tube contains a mixture of helium and neon.
The helium is excited by electric discharge. The excited state is a triplet state (which
therefore has a long lifetime) and the energy of the excited helium is transferred to
the neon, which decays in two stages, one of which is responsible for the laser action.

15 Radioactivity

15.1 Electron volt

The energy of radioactive particles is often expressed in electron-volts (eV) rather than the SI unit (joule). 1 eV is the energy gained by an electron when it is accelerated through a potential difference of 1 volt, and equals $1·6 \times 10^{-19}$ J. The energy is usually of the order of one MeV (= 10^6 eV).

15.2 The nucleus and its radiation

A characteristic property of many nuclei is that they are unstable, and this instability results in the nucleus emitting radiation. There are many forms of radiation. As biologists we are interested in just three, known as α, β and γ rays. In order to understand what happens to the nucleus when any of these forms of radiation is emitted it is necessary that we have a very elementary idea of the make-up of the nucleus.

Nuclei are composed of nuclear particles, of which the two heaviest are the proton (p) and the neutron (n). The mass of the neutron is equal to the sum of the masses of a proton and an electron, and it is often convenient to consider the neutron to be made up of one proton and one electron. The main difference between the proton and the neutron is that the proton is a charged particle, with one unit of positive elementary charge, whereas the neutron is uncharged.

Any individual nucleus will contain an integral number of protons (Z) and neutrons (N). The atomic number, which specifies the chemical species pertaining to that nucleus, is equal to the number of protons. A number of different forms of each chemical species is found, differing in the number of neutrons contained in the nucleus. These are known as different *isotopes* of that particular chemical. Thus, for example, hydrogen can be obtained with 0, 1 or 2 neutrons (but always with 1 proton, otherwise it would not be hydrogen). The different forms of hydrogen are given different names (deuterium has 1n and tritium has 2n), but for no other chemical are the different isotopes distinguished by separate names. In order to distinguish the different isotopes it is usual to write their atomic number (Z) as a subscript, and the sum of their protons and neutrons $(N + Z)$ as a superscript. Thus deuterium is $_1H^2$ and tritium is $_1H^3$.

We saw in Chapter 10 that strong repulsive electrostatic forces exist between like charges. In order to overcome the electrostatic forces in a nucleus due to the presence of protons there are extremely strong attractive nuclear forces tending to keep the nucleus intact. Any nucleus can exist in one of a number of discrete energy levels. These energy levels do not have the simple-minded interpretation of the discrete energy levels available to the electrons surrounding the nucleus (in terms of separate orbits), but they are quantized in the same way, and the nucleus can jump from one to another, absorbing energy in the upward transition and releasing energy, in the form of electromagnetic radiation in the downward transition. As before the frequency of the emitted radiation will be given by

$$\Delta E = h\omega$$

where ΔE is the energy gap between two energy levels, which for nuclei is much greater than that for the electronic energy levels, with the consequence that the frequency of the emitted radiation is very high. Such radiation is known as γ rays.

Not all nuclei are stable. In order to be stable a nucleus must contain the correct proportion of neutrons and protons. For nuclei of low atomic number this is obtained when the proportions are about 1p : 1n; at high atomic number the ratio is about 1p : 1·5n. If a particular nucleus contains an excess of either neutrons or protons, then it will try and redress the balance by changing the proportions by emitting particles of one type or another.

If there are too many protons in the nucleus compared with the number of neutrons, then the nucleus will either emit an α particle (which is a helium containing 1n and 2p — i.e. $_2He^4$), or else a particle known as a positron which is identical to an electron but with a positive charge, and usually written as a β^+ particle. If the nucleus emits an α particle, then the new nucleus contains two less protons and two less neutrons; after emitting a β^+ particle the new nucleus contains one less proton, but one more neutron — the proton has in effect been converted to a neutron, releasing its positive charge as a positron. Fig. 15.1 illustrates the nature of these transitions. The different possible nuclei of the first twelve elements are denoted by symbols showing their mode of decay, and a few transitions are indicated by thick lines. The line through $N = Z$ has also been drawn.

At low atomic numbers the release of α particles is uncommon, and nuclei containing excess positive charge for their neutron content radiate positrons. At high atomic number both particles and β^+ particle decay is found.

There is a strong attractive electrostatic force between positrons and electrons, and thus the positron is a short lived particle, and travels only small distances before interacting with an electron. The interaction causes the annihilation of both particles with the release of γ rays. The energy of the γ rays so released is always the same and equal to 0·51 MeV. Therefore the radiation observed following β^+ emission is γ radiation.

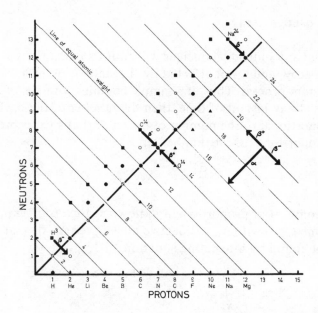

Fig. 15.1 Nuclear chart of first 12 elements. ● = major naturally occuring isotope; ○ = other stable isotopes; ■ = β^- emitters; ▲ = β^+ emitters. There are no α particle emitters at these low atomic numbers.

If there are too few protons in the nucleus for the number of neutrons present, then the nucleus will emit an electron (known as β^- radiation, or often just as β radiation); the effect on the nucleus will be to increase its proton number by 1, and to reduce the number of neutrons present by 1. Such a transition is also shown in Fig. 15.1

After a nuclear particle has been emitted from a nucleus the new nucleus formed will not necessarily be in its lowest energy (or ground) state. If it is left in a higher energy level, then the nucleus will spontaneously jump to its ground state, releasing γ rays as discussed above. Thus, associated with any of the forms of radiation there may also be γ radiation (e.g. see Fe^{59} in Table 15.1.).

The total energy released by any form of radiation emission will be characteristic of the particular transition. α particles and γ radiation are always released with the precise energy characteristic of the decay. For α particles, this is generally in the region of several MeV. β particles, however, are always emitted together with another nuclear particle, known as a neutrino, and the energy change of the transformation can be apportioned out between the β particle and the neutrino in all different proportions. Thus the energy of the β radiation can take any value up to a maximal level, obtained when the energy of the neutrino is zero. The maximal energy of β particle decay is usually not greater than about 2 MeV.

15.3   Law of radioactive decay

The probability that any individual nucleus will emit its radiation is a characteristic
of that particular decay, and is not affected by the state of other nuclei in the sample.
The decay is a random process. Thus, if a sample contains N nuclei of a particular
radioactive species at any instant of time, then the number of these which will decay
in unit time is proportional to the number of nuclei of that species present; i.e. the
rate of decay is proportional to the number left:

$$\frac{dN}{dt} = -\lambda N \tag{15.1}$$

in which λ is the constant or proportionality known as the *disintegration constant*.

If we start at time $t = 0$ with N_0 radioactive nuclei present, then at time t the
number left will be found by integrating equation 15.1, giving

$$\int_{N_0}^{N_t} \frac{1}{N} \, dN = -\int_0^t \lambda \, dt \tag{15.2}$$

i.e. $$\ln(N_t/N_0) = -\lambda t \tag{15.3}$$

We can make use of the relationship that if $\ln(\alpha) = x$ then $\exp(x) = \alpha$ to convert
equation 15.3:

$$N_t = N_0 \exp(-\lambda t) \tag{15.4}$$

Thus the radioactivity of any particular sample of radioactive material, measured in
terms of disintegrations per unit time, follows an exponential decay, with a rate of
decay governed by the disintegration constant λ, Fig. 15.2. It is common to talk
about the half life of a radioactive material. This is the time taken for the radioactivity to fall to half its value. Thus in equation 15.3, writing $N_t = 0.5 \, N_0$ gives:

$$t_{\frac{1}{2}} = \ln(0.5 N_0/N_0)/\lambda = 0.693/\lambda \tag{15.5}$$

Note that the half-life is the time taken to decay from the value of N at any time, to
half that value one half-lifetime later. In other words, if we start at time zero with
N_0 radioactive nuclei present, then after $t_{\frac{1}{2}}$ seconds there will be $N_0/2$ such nuclei
left, after a further $t_{\frac{1}{2}}$ seconds there will be $N_0/4$ nuclei left and so on.

The properties of some radioactive nuclides commonly used are given in Table
15.1.

15.4   Unit of radioactivity (curie)

The activity of a radioactive sample can be measured in terms of the number of
disintegrations occuring per second. The unit of radioactive activity is the *curie*,
abbreviated Ci, which is defined as 3.7×10^{10} disintegrations per second (d.p.s.).

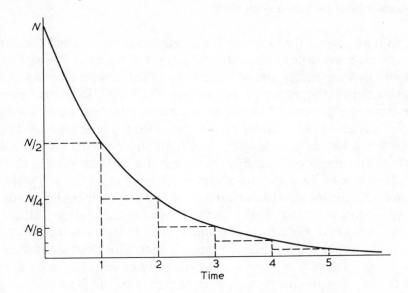

Fig. 15.2 Exponential timecourse of radioactive decay.

TABLE 15.1

Radioactive elements frequently used by biologists

Nuclide	Radiation	Energy (MeV)	Half-life
H^3	β^-	0·018	12·3 y
C^{14}	β^-	0·156	5770 y
Na^{22}	β^+	0·54 (→ 0·51 A.R.) ⎫	2·58 y
	γ	1·277 ⎭	
P^{32}	β^-	1·71	14·2 d
S^{35}	β^-	0·167	87 d
K^{40}	β^-	1·33 ⎫	1·25 × 10^{9}y
	γ	⎭	
Ca^{45}	β^-	0·254	167 d
Fe^{59}	β^-	1·56	45·1 d
	γ	1·1, (56%) ⎫ 1·29 (44%) ⎭	

(→ 0·51 A.R. = annihilation radiation produced when positron meets an electron. γ radiation of 0·51 MeV is the result.)

This is approximately equal to the activity of 1 gm of pure radium, and is a very high activity; a more useful unit is a biology department is a μCi (3·7 × 10^4 d.p.s.).

15.5   Interaction of radiation with matter

Both the ability to detect the radiation from radioactive decays, and the biological effects of the radiation depend upon the interaction of the radiation with matter.

α particles lose their energy almost entirely by colliding with electrons, causing the ionization of the atom to which the electron was bound. Ionization energies are of the order of a few tens of eV, whereas α particle energies are of the order of a few MeV. Thus about 10^5 collisions are required to stop an α particle. The distance travelled by the α particle in making these collisions depends upon the density of electrons in the material. Very roughly, the range of α particles in air is about 1 cm for each 1 MeV of initial energy, and is very much less in water or any solid.

β particles with energy less than about 1 MeV too lose their energy almost entirely by colliding with electrons, with the consequent ionization of the material. The differences between the effects of β and α particles is due to the difference in mass of the two particles. β particles are much lighter. This means that their direction is changed by much greater amounts during the collision (due to the conservation of momentum), and it also means that their velocities are much greater for a given energy, with the result that they travel further between collisions. Roughly speaking electrons travel 100 times as far in a material as α particles. Very approximately, electrons with energies in the range 0·5—10 MeV will travel 2—3 metres in air per MeV of initial energy, about 3—5 mm/MeV in water and about 500 μm/MeV in lead.

β particles with energies greater than about 1 MeV also lose energy by a process known by *bremsstrahlung*, in which the kinetic energy of the electron is converted to γ radiation due to interaction between the nuclei of the material and the electrons. The contribution to energy loss by this means increases rapidly for electrons of increasing energy. Although the β radiation itself is absorbed gradually by the creation of γ rays, the γ rays themselves are far more penetrating than the original β particles. Thus, whereas virtually all the energy of the β particles of initial energy less than 1 MeV is absorbed as described above, that contribution of the energy of more energetic β particles that is converted to bremsstrahlung requires much greater distances for its absorption, as discussed under γ radiation.

γ radiation interacts with matter in three ways, known as 1. Photoelectric absorption. The γ ray knocks an orbital electron from its atom, disappearing in the process (i.e. all the energy of the γ ray is transferred to the electron). This electron will lose its energy as discussed above for β particles. 2. Compton scattering. The γ ray is scattered by an electron, losing some energy to the electron. 3. Pair production. Provided the γ ray initially has an energy greater than 1·02 MeV, then the energy of the γ ray can be converted into an electron plus a positron, which then lose their energies as discussed above. Of these three types of interaction, photoelectric absorption is the most probable at low energies (less than about 0·5 MeV) and with materials of high atomic number, Compton scattering at intermediate energies and with materials of low atomic number and pair production at high energies.

One difference between absorption of γ rays and absorption of α or β particles, is that, whereas α and β particles are completely absorbed by materials above a certain thickness for a given initial energy, γ rays obey an exponential absorption law, and are never completely absorbed. Whereas the energies of α and β particles are lost gradually in a series of many thousands of collisions, the energy of a γ ray is usually lost completely in a single process, and thus the energy of the γ rays leaving a piece of material is the same as that entering it — the absorption means that there is a smaller total number, not a smaller energy per individual γ ray. (This is not the case with Compton scattering, and the complexities introduced by this type of interaction mean that the absorption of γ rays is only approximately given by the exponential law.)

The law showing the effectiveness of a material in absorbing γ rays is

$$I = I_0 \exp(-\mu x) \tag{15.6}$$

where I is the intensity after passing through a thickness x of the material, with an input intensity I_0. μ is known as the *absorption coefficient*. The value of μ is determined both by the nature of the absorbing material, and by the energy of the γ rays. Very approximately, for water, μ varies from about $150\,\mathrm{m}^{-1}$ for γ rays of 0·1 MeV to $3\,\mathrm{m}^{-1}$ at energies of 10 MeV. The values in lead are about ten times larger.

15.6   Detection of radioactivity

Notice that the end result of the interaction of radioactivity with matter is most often an eventual ionization of the material absorbing the radiation. The three common methods of detecting the radioactivity make use of this ionization.

1. Photographic film. The ionization of the emulsion of the photographic film results in a blackening of the film. People working with radioactive materials commonly wear a badge consisting of a sheet of photographic material. In order to estimate the different types of radiation, different areas of the film can be protected by filters — comparison of the blackening in the different areas allows both the amount and type of radiation recieved by the badge, and thus of the individual, to be monitored.
2. Geiger-Muller tube. This is a tube, filled with gas at low pressure (about 1/100 atmospheric), whose outer wall forms one electrode, and which contains an internal electrode in the form of a wire. If a sufficiently high voltage is applied between the two electrodes then the gas will ionize. If a slightly lower voltage is applied across the tube, then the gas is not ionized and there is no electrical continuity between the two electrodes. A radioactive particle entering the tube can now induce ionization, enabling current to flow between the two electrodes, and thus enabling the radioactive particle to be detected.
3. Scintillation counter. If the ionized material formed by interaction with the

radioactive material is left it will rapidly recapture an electron, and in so doing will emit a photon of light at a wavelength determined by the energy level of the electron knocked out of the atom during the ionization procedure. The wavelength of the photon released will often be in the visible part of the spectrum, and can be detected by a sensitive light detector such as a photomultiplier. The intensity of the light flash caused by any particular incoming radiation will be proportional to the number of ionizations caused by the radiation, which will be dependent upon the input energy of the radiation. Thus the scintillation counter gives information about both the timing and the energy of the incoming radiation, and it is possible for a single counter to distinguish between different energies of incoming radiation, and thus between radiation from different nuclei, e.g. P^{32} and C^{14}. The liquid scintillation counter is the standard instrument used for measuring radioactivity in biochemical experiments. In this the radioactive sample is placed in a bottle together with the scintillant (a chemical particularly efficient at producing light of the correct wavelength for the photomultiplier), dissolved in a suitable solvent (usually toluene). The sample will preferably be in liquid form, but solid samples are acceptable. This method of use permits the absorbing material (the scintillant) to be in as close a proximity as possible to the source of radiation, and thus the counting efficiency (the fraction of the radioactive particles emitted which are actually detected) is as high as possible. With the Geiger-Muller counter for example the sample is always removed from the counter by at best a short air gap and a mica window.

One handicap of the scintillation counter is that water is a very poor solvent because it quenches the light produced by the scintillant very effectively; moreover, water is immiscible in toluene, which is the best solvent. Although there are methods of obviating this problem to a large extent, the absence of any water is best and this can be an obvious drawback when detecting radioactivity in biological preparations.

15.7   The biological use of radioactivity

Tracers (labelling)

1. Location in space (autoradiography). The whereabouts of a chemical within a cell or within the body can be followed by labelling the chemical with a suitable radioactive nuclide, and then locating the position of the radioactivity. This can be done microscopically (*autoradiography*) under the light or electron microscopes by coating the relevant sections with a thin layer of photographic film, leaving for sufficient time for the radioactivity to expose the film and then developing the film. The radioactivity will show up as black spots. Elsewhere the photographic film will be transparent, and the details of the section can be seen for identification of the location of the radioactivity.

Location of radioactivity can also be done macroscopically; to study for example the location of a particular chemical in the body. The localization can be made by positioning counters around the body able to accept radiation from a precise direction only. The radiation travels in straight lines, and thus location can be identified.

2. Location with respect to time. The extent of a reaction can be followed by labelling the reactants in such a way that one of the products which can be separated from the reactants, is labelled. For example, the terminal phosphate of ATP is often made radioactive (using P^{32}). It is possible to separate phosphate from ATP, and thus to determine the extent of hydrolysis of ATP by effecting the separation and measuring the amount of phosphate present by measuring the intensity of the radioactivity.

The amount of a substance transported across a membrane can be determined for example by labelling that substance with a suitable radioactive atom (C^{14} is often suitable, and can be readily introduced into most organic substances), and measuring the radioactivity on whichever side of the membrane is desirable. Thus, if the uptake of an amino-acid by red blood cells is to be measured, the red blood cells can be added to a medium containing the labelled amino acid. After a suitable length of time the blood cells are removed (by centrifugation) and the radioactivity present in the blood cells then measured.

3. Location in sequence. The sequence of steps in a series of biochemical reactions can be followed by labelling particular reactants and investigating their fate during the reactions. For example, the study of the processes of protein synthesis has frequently made use of the ability to label individual amino-acids and investigate to which constituents of the cells they are attached at various stages of the synthesis.

Isotope dilution

This is a method for estimating numbers in populations, volumes of fluid in a body etc. The process involves obtaining a sample from the population/volume, labelling this in some way and reintroducing the labelled sample to the source. After sufficient time to allow thorough mixing of the labelled and unlabelled portions another sample is taken and the level of radioactivity in this sample measured. An estimate of the total size can then be obtained by simple proportionality. For example, 1000 ants are taken from an ants nest, fed on labelled sugar and reintroduced. Some time later a sample of 750 ants is taken from the nest and the number of 'labelled' ants in this sample is found to be 125. The total population size is thus about

$$\frac{750}{125} \times 1000 = 6000.$$

Statistical tests must be applied to tell the accuracy of the preduction.

Note. For the above uses it is desirable to use radioactive nuclei with long half-lives compared with the duration of the experiment.

Radioactive dating

The best known method of dating of fossil material is by means of C^{14} dating. The CO_2 in the atmosphere contains a certain percentage of C^{14}. This arises from the action of cosmic radiation on the nitrogen

$$_7N^{14} + \beta^- \rightarrow {}_6C^{14} \tag{15.7}$$

The carbon in living animals will contain the same proportion of $C^{14} : C^{12}$ as the atmosphere. Upon death the remains of the animal which will eventually become fossilized thus contain a certain ratio of C^{14} to C^{12}. The C^{14} will decay, with a half-life of 5770 years. A negligible amount of the nitrogen present will be acted upon by cosmic radiation and thus be converted to C^{14}. Thus the proportion of C^{14}/C^{12} in the fossil will change with time, and the observed ratio used to date the fossil.

In order for this method to work the proportion of C^{14} in the atmosphere at the time of death of the animal must be known. Until recently it was assumed that this has remained constant for the last fifty thousand years or so (at $10^{-8}\%$ C^{14}), the limiting age testable by this method. Recently a method of testing this assumption has been found. The age of trees can be determined accurately by their annual rings. Bristle-cone pine trees in California can be found covering a time span of the last 10 000 years or so, and thus samples of known age can be obtained and used to calibrate the C^{14}-dating technique. Thus over this period of time an accurate calibration is known; the assumption that the present day ratio has remained constant no longer holds for age greater than about 3000 years.

15.8　Biological effects of radiation

The biological effects of radiation arise from the ionization produced and the effects of this on the tissues and molecules of the body. The action can either be direct – an ionization of a biological molecule – or indirect via the ionization of water

$$H_2O \rightarrow H_2O^+ + \text{electron } (e^-)$$

The free electron may then combine with another water giving

$$H_2O + e^- \rightarrow H_2O^-$$

Both H_2O^+ and H_2O^- are unstable and dissociate rapidly to give rise to a free radical

$$H_2O^- \rightarrow H^* + OH^-$$

$$H_2O^+ \rightarrow H^+ + OH^*$$

Where H^* and OH^* are the highly reactive free radicals. These either interact with water, or with themselves, or else with any other molecule in the vicinity, which in the living cell will often be a biological macromolecule.

The ionization has a high probability of eventually causing the breaking of a bond in a macromolecule, which in most cases will cause that molecule to become biologically ineffective. The damage this causes to the cell obviously depends upon the importance and replaceability of the molecule. Disruption of the chromosomes produces the most lethal effects, and the sensitivity to radiation is greatest in the rapidly dividing cells of the body — the gonads, bone marrow, intestinal epithelium and the skin.

The effect of any incoming radiation will thus depend upon the amount of ionization that it is able to cause. We saw earlier that this was dependent upon the energy of the radiation, the major fraction of the energy of the incoming radiation being lost by collisions between the radiation and electrons in the material. However the damage caused to a cell will also be dependent upon the density of the ionization, and for this reason particles with low velocity (such as α particles) will have a greater effect upon the tissue than particles with high velocity (e.g. β particles) since the density of ionization, as we have seen, is greater the lower the velocity.

Radiation dosage is measured in two ways 1. in terms of the *energy absorbed per unit mass*, irrespective of the type of particle causing the radiation (and thus the density of ionization along the track of the incoming radiation) and 2. taking account of the importance of this effect.

The *rad* (for *r*adiation *a*bsorbed *d*ose) is the unit of radiation dosage taking account of the energy absorbed alone, and is an energy absorption of 10^{-2} J/kg which is about equal to 6×10^7 MeV/g. (An earlier, and less biologically useful unit, is the *roentgen*, defined in terms of the ionization caused in air by the radiation. Its value is equivalent to about $9 \cdot 3 \times 10^{-3}$ J/kg, and is thus virtually equal to the rad.)

In order to take account of the effectiveness of the particular type of radiation, each form of radiation is given a relative biological effectiveness (*RBE*) factor, which is determined from experiments investigating the effect of different types of radiation on a variety of biological preparations. γ rays and β particles have an RBE of 1, and α particles an RBE of about 10 (altogether some sources quote higher values).

The unit of dosage accounting for the RBE factor is the *rem* (*r*oentgen *e*quivalent in *m*an), which is the dose in rads multiplied by the RBE factor.

$$\text{rem} = \text{rad} \times \text{RBE} \tag{15.8}$$

In order to evaluate the dosage received from any radioactive source we need to know the energy of the radiation from the source of known activity, the type of radiation, the fraction of the radiation absorbed by the body and the time of exposure.

Example 15.1

You are working with a compound containing radioactive C^{14}, and are foolish enough to pipette a quantity of this by mouth and accidentally swallow 10μCi. What is your radiation dosage per minute, assuming the C^{14} becomes distributed throughout your body?

The energy of the β^- particles (RBE = 1) from C^{14} (see table 15.1) is 0·155 MeV. Their range in water is sufficiently small that they will all be absorbed by the body. Let us assume that your mass is 50 kg. The rate of disintegration from a source of $10\mu Ci$ is $(3\cdot 7 \times 10^{10}) \times (10^{-5})$ disintegrations per second thus the dosage is

$$3\cdot 7 \times 10^{10} \times 10^{-5} \frac{\text{disintegrations}}{\text{second}} \times \frac{(60 \text{ second})}{\text{minute}} \times \frac{0\cdot 155 \text{ MeV}}{\text{disintegration}}$$

$$\times \frac{\text{g rad}}{6 \times 10^7 \text{ MeV}} \times \frac{1}{50 \times 10^3 \text{ g}}$$

$$= \frac{3\cdot 7 \times 10^{10} \times 10^{-5} \times 60 \times 0\cdot 155}{6 \times 10^7 \times 50 \times 10^3} \frac{\text{disintegrations} \times \text{s} \times \text{MeV} \times \text{g} \times \text{rad}}{\text{s} \times \text{min} \times \text{disintegration} \times \text{MeV} \times \text{g}}$$

$$= 1\cdot 15 \times 10^{-6} \text{ rad/min}$$

Since the RBE factor is 1 this gives a dosage of about 1 μrem/min.

This calculation assumed that the radiation was distributed evenly throughout the body. Of course this will not usually be the case. Before it is absorbed we might consider that a more appropriate estimate of dosage would be to consider that all the radiation was absorbed by 100 g of your intestine, in which case the dosage received by your intestine (a part of the body particularly sensitive to radiation) will be 500 μrem/min, i.e. 0·5 mrem/min.

Notice that with the C^{14} kept in the pipette externally your dosage will be very much smaller, for several reasons. In the first place there will always be considerable absorbing material for β particles between your body and the radioactivity, and secondly you will only receive a fraction of the radioactive disintegrations, since these will have random direction.

15.9 Maximum permissible doses

The numbers in the above calculation have little meaning without an understanding of the magnitude of doses causing particular effects. Obviously any dose at all can have some effect, and to some extent the line laid down for a maximal permissible dose is arbitrary. In order to put the figures in perspective we shall discuss the maximal permissible doses laid down by the International Commission on Radiological Protection in relation to some other figures.

The total radiation dose that you receive per year from natural sources (cosmic radiation, natural terrestrial sources and naturally occuring isotopes in your body) is about 100 mrad/year. Thus, in calculation 15.1 above, the C^{14}, evenly distributed in the body, is increasing your background radiation input by a factor of about 5.

It is estimated that an average exposure of 50 rem per person per lifetime is necessary to double the mutation rate, or alternatively that an average exposure of

TABLE 15.2

Part of body	Maximum permissible dose (rem)	
	In any 13 weeks	In any year
Gonads, bone marrow, whole body	3	5
Skin, bone, thyroid	8	30
Any other single organ	8	15
Hands, forearms, feet, ankles	20	75

1 rem/generation will increase the mutation rate by 1/70.

At the high end of the scale a single exposure of about 250 rads is the dose required to produce a 50% fatality in man in 30 days. This is due to damage to the bone marrow, preventing replacement of the blood cells. Higher doses produce death more quickly, single doses above about 1000 rads preventing replacement of the gastro-intestinal epithelium and causing death in a few days, and doses above 10 000 rads causing death due to disruption of the central nervous system in a matter of hours. Doses smaller than 200 rads, though not causing short-term death, shorten the life-span of the animal and induce cancer. The effect of a given dose is smaller if the dose is accumulated over a period of time, rather than in a single exposure.

Table 15.2 lists the maximal permissible doses for various parts of the body for people who are occupationally exposed. The figures for the general public are one-tenth these values.

15.10   Questions

1. Show that there are 1.6×10^{-19} J/eV.
2. The half-life of Na^{24} is 15 hours. What is its disintegration constant? How long before a given sample has decreased in strength by a factor of 10?
3. A radioactive sample initially records 42 000 d.p.s. After 1 day the level has fallen to 12 000 d.p.s. What is the half-life of the radioactive species present?
4. You are provided with 1g of solid ATP (M.Wt, 551) in which some of the terminal phosphates are labelled with P^{32}. If the strength of the sample is 100 μCi, what fraction of the ATP molecules are labelled?
5. What percentage of the γ radiation from Na^{22} is absorbed by a sheet of lead (a) 1mm, (b) 5mm thick?
6. You swallow the entire sample of question 4. What is the dose rate assuming that the radioactivity is evenly distributed in your body? If the entire energy is absorbed by your stomach, what dose rate is now relevant? How long before your 13 week maximal permissible dose as given in Table 15.2 is reached, assuming that the dose

rate remains constant, and that the relevant mass of your stomach is 100 g? Moral: don't pipette radioactive samples by mouth.

7. A Carboniferous fossil, weighing 10 g, is 90% carbon and has a radioactive decay of 36 d.p.s. Assuming that the proportion of C^{14} in the atmosphere has remained constant, how old is the fossil?

16 Control

A very common phenomenon in biological systems is that of control of some output or other. A few examples will illustrate the problem.

1. Within any cell the correct concentration of the various constituents is maintained at a more-or-less constant level. For example, in the most active tissue known — the flight muscle of large insects — the concentration of ATP does not change to any marked degree even through the change in usage of ATP between rest and activity changes by a factor of a thousand times or more. This requires the cell to be able to sense when ATP is being used, and to use this means of sensing to switch on the production of ATP at the correct time so as to synthesize the correct amount. In principle of course it would be possible for the signal for the muscle to become active (a change in the calcium ion concentration within the cell) also to trigger the production of ATP. Indeed this does happen. But the muscle can use ATP at different rates even with the same calcium ion concentration (due to the different loading on the muscle under different conditions), and to maintain the correct concentration requires some other method of control. In fact the cell has several methods. All the subsidiary methods, however, rely on the production of ATP from the energy of the fuel. Different insects use different fuels (carbohydrates or fat) so we will not be too specific. There is one particular control enzyme in the pathway for the breakdown of this fuel — phosphofructokinase (PFK). This enzyme is activated by ADP and AMP and inhibited by ATP. Thus when the ATP in the muscle is broken down (resulting in more ADP and AMP and less ATP) the enzyme is activated and so more ATP is produced. When the correct levels are restored the enzyme is inhibited. By this means the correct concentrations can be controlled at all times.

2. It is essential for an animal to be able to position its limbs correctly, and to be able to keep them at the required position even when the load on that limb changes. This requires the animal to have some means of sensing its limb position, so that corrections can be applied when the limb moves from the position it is required to have. We use at least two systems to do this. The first is our visual system, whose operation in this respect is straightforward and under voluntary control. The other method is not completely under voluntary control and is known as the muscle

spindle system. Within our muscles are a number of specialized cells known as muscle spindles. We shall look at these in more detail later, but essentially they are sensory fibres which respond to muscle length. Thus when the muscle is extended the sensory nerves leading from the muscle spindle become more active, and when the muscle becomes shorter these nerves become less active. The nerves connect directly to the motor nerves whose activity controls the contraction of the muscle. Thus, if the muscle is extended (because a load is applied to it for example), the activity of the sensory fibres increases, thereby increasing the activity of the motor fibres and thus of the muscle. The muscle therefore contracts to restore its original length. An equilibrium is maintained with a balance between extension in the muscle and activity of the muscle. Anything which tends to upset this equilibrium causes the muscle to change its activity so as to redress the imbalance.

3. As a final example we will consider the control of wing-beat frequency in some insects. Most orders of insects (but not all — see page 178) beat their wings at a frequency determined by the frequency of arrival of nerve impulses to the flight muscles. The relevant nerves originate in the thoracic ganglion. In locusts, and probably in other insects, the frequency of impulses in these nerves is dependent upon the overall degree of activity of sensory input to the ganglion, one important contribution to which arises from stretch receptors at the base of the insect wing. The larger the amplitude of movement in the wing, the more these stretch receptors are excited, and thus the greater the sensory input to the ganglion, and the higher the frequency of nerve impulses to the muscle, and thus the higher the wing-beat frequency. Now because of the inertia of the wings, the higher the wing-beat frequency, the smaller the amplitude of the movement of the wings, and thus the smaller the stimulation of the stretch receptors. We can see that an equilibrium will be set up at which the frequency of wing-beats gives rise to that amplitude of movement of the wings that is just able to provide sufficient excitation to cause that frequency to be maintained. If for some reason, the output frequency drops, then this results in more excitation due to the increased amplitude of movement of the wings, and an increase in the wing-beat frequency again. Likewise, if the frequency increases by some means, the excitation from the stretch receptors will be lowered, resulting in a lowering of the frequency.

All these, and other, control systems have a number of properties in common. There are also a number of problems associated with maintaining a stable output. It is with these properties, and problems, that this chapter is concerned.

Fig. 16.1 shows diagrammatically the general features of a control system. The essential features are that some measure of the output is fed back through a *feedback* pathway, and compared with the input to the system. The difference between input and output is then used as the signal which drives the effector which controls the output. Such control is often known as *servo-control*. The particular ways in which the three examples fit into this framework are illustrated in Figs. 16.2 and 16.4.

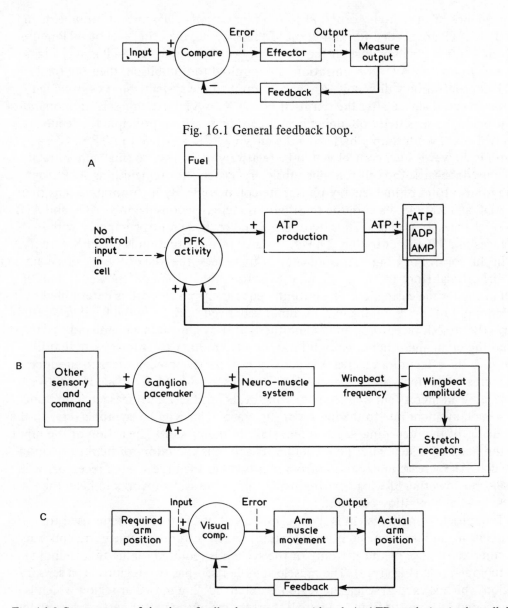

Fig. 16.1 General feedback loop.

Fig. 16.2 Components of the three feedback systems considered. A. ATP regulation in the cell. B. Control of wing-beat frequency in insects. C. Control of arm position by the visual system.

For the maintenance of ATP levels in the cell, the output is obviously the ATP concentration. The comparison between the required ATP concentration and the actual concentration is made by the enzyme PFK. In this situation there is no obvious input to the system in the cell. Any particular muscle cell has a particular level of ATP that is to be maintained, and this is obtained by having a particular dependence of the activity of the PFK as a function of ATP concentration. If high levels of ATP are needed for inhibition of the enzyme, then the level of ATP maintained in the cell will be high; if the sensitivity of the enzyme to ATP is very high, whereby only low ATP concentrations are required for inhibition, then the final ATP concentration will be low. An overall input to the system could be provided by having a substance alter the sensitivity of PFK to ATP. Natural selection operates to provide the sensitivity obtaining in any particular cell. In principle the feedback control could work simply by having the enzyme PFK sensitive to ATP concentrations; however such control would be relatively insensitive to small amounts of ATP used, because the cell contains subsidiary systems for replenishing ATP from temporary stores of high-energy phosphate compounds. By incorporating sensitivity to ADP and AMP levels also, the sensitivity is increased considerably. ADP and AMP levels changes as soon as ATP is hydrolysed and by large ratios compared with changes in ATP concentration. The error signal in this situation is the PFK activity.

In the control of wing-beat frequency in the insect, the role of the various elements is fairly straightforward. This system uses as its comparator a pacemaker system or cell in the thoracic ganglion. The output frequency from this cell is determined by the amount of input coming in. This input is derived both from the feedback pathway (the stretch receptors), and from other pathways, including command signals from the brain and other sensory information from other receptors. Note that in this case the effect of an increase in activity in the stretch receptors is to produce an increase in activity in the pacemaker. The reason why the feedback is negative is because an increase in frequency of wing-beats produces a decrease in amplitude of wing movement due to the inertia of the wings. The + and − symbols refer to the influence of the preceding stage on that stage in the process. The action of the input to the pacemaker depends upon whether its action is excitatory or inhibitory upon the cell. An increase in excitation from inputs other than the stretch receptors will result in a lowering of wing-beat frequency, since the system works to keep the total level of excitation constant.

The muscle spindle system is the most involved of the three systems used for illustration, and has some additional features which introduce some extra concepts. Further knowledge of the anatomy of the system is required for an understanding of the operation (Fig. 16.3). The muscle spindle itself has two regions: 1. a sensory region which is excited when the region is extended; 2. a muscular region which is innervated by a motor fibre and which shortens when the motor nerve is stimulated by an amount proportional to the frequency of stimulation. The motor fibres innervating the muscle region of the spindle are known as the γ fibres.

The total length of the spindle is the same as that of the main body of the muscle, since both are connected to the same tendon. Length changes in the sensory region can be caused either by a change in the overall length of the entire muscle, or by an excitation of the muscular region of the spindle while the length of the muscle is constant. It is this latter feature that provides the input to the system. The feedback

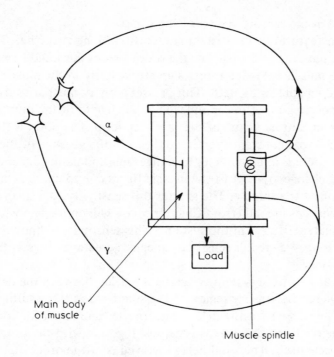

Fig. 16.3 The muscle spindle system.

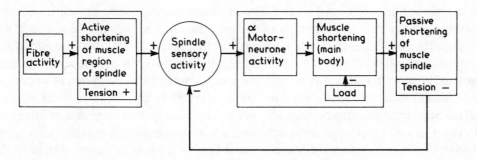

Fig. 16.4 Block diagram of the muscle-spindle system.

loop uses the output from the muscle spindle sensory region to maintain the activity to the extrafusal muscle fibres. If the load on the muscle changes, thereby extending the muscle, this is detected by the spindle and fed back as an increased activity, thereby causing the muscle to contract. The error signal is the output from the sensory region of the spindle. The equilibrium length of the muscle is determined by the length of the sensory region of the spindle. If this is changed then the feedback corrects it. Thus if the length of the muscular region of the spindle is changed (by activity in the γ fibres) the length of the whole muscle will change to correct the extension caused in the sensory region of the spindle.

Two particular features of the system are worth looking at further. The first concerns the magnitude of the error signal (the output from the spindle sensory region). Obviously, if the muscle is loaded more, a greater activity in the muscle will be required to maintain a constant length. This greater activity must arise from a lengthening of the sensory region of the spindle. Thus the length of the muscle will not be held absolutely constant by virtue of the feedback loop. The point is that the length will be held much more constant than it would be in the absence of the feedback. How much error there is in actual length changes due to loading depends upon the sensitivity of the change in input to the muscle fibres due to changes in length of the sensory region of the spindle. The greater this sensitivity, the smaller the length changes due to changes in load (from Chapter 5 you will appreciate why this is described as a greater stiffness in the system.) This sensitivity is known as the *gain* of the feedback loop — a greater sensitivity means a greater change in frequency for a given length change.

The second feature of the system concerns the delay between the detection of the length change by the sensory region of the spindle, and the resulting movement of the muscle. The reasons for the delay concern the time taken for conduction of impulses down nerve fibres and across synapses, together with the process of excitation of the muscle itself. The total delay involved is of the order of 100 ms. The effect of the delay is to introduce the possibility of oscillations into the system. Imagine what happens with a muscle supporting a load. If there is an extension of the muscle for some reason, this is detected and an increased activity in the muscle causes a shortening. The load will be lifted towards the equilibrium point, and both because of the inertia of the load, and because the muscles will still be active it will overshoot the equilibrium. The overshoot will be detected, but it will be about 100 ms before any action is taken in terms of reduction in activity in the muscle. The same problems arise on the way down, and the result is to cause oscillations to occur in the system. These are usually most unsatisfactory. They can be quite easily demonstrated. Support your arm on a bench and try to point your finger at a pointer keeping it as still as possible. You will notice your finger vibrating slightly. A simple photocell/light arrangement will easily show the oscillations, which can then be displayed on a penrecorder. The frequency of oscillations will be about 10 Hz.

The way in which such oscillations can be minimized is by feeding back not only information about the position or length of the muscle, but also about its velocity, and preferably about its acceleration also. This enables predictions to be made about the future length of the system, and thus to counteract the effects of the delays. Thus, in the system during the oscillations, notice that the velocity is greatest when the length is passing through its equilibrium point. Information about the velocity thus precedes information about the length in terms of required correction by one quarter of a period of the oscillation.

In our generalized system, let θ_i be the magnitude of the input, and θ_o the magnitude of the output. The feedback circuit can be made to amplify the output, and if this amplification is β, then the error signal (e) is $\theta_i - \beta\theta_o$. The effector circuit can also amplify the error signal, and if this amplification is A then it follows that

$$\theta_o = A(\theta_i - \beta\theta_o)$$

$$\theta_o + A\beta\theta_o = A\theta_i$$

Rearranging

$$\theta_o(1 + A\beta) = A\theta_i$$

$$\theta_o = \frac{A\theta_i}{1 + A\beta}$$

Provided that $A\beta$ is much greater than 1 (which can easily be arranged in principle by making A large enough), then

$$\theta_o \doteq \frac{A\theta_i}{A\beta} = \frac{\theta_i}{\beta}$$

Thus the output is determined by the nature of the feedback loop, and is independent of the nature of the effector. This, mathematically is the reason why a very high gain in the system gives rise to a high degree of stiffness in the muscle spindle system. The output follows the input, independent of any loading.

The velocity and acceleration feedback can also be treated mathematically, but more advanced mathematics is required, and is beyond the scope of this book.

It is very often the case that the gain of the system is high; this means that investigating the operation of the system in its normal form is difficult. Thus one would get relatively little information about the way the muscle spindle system operates by keeping the system intact and measuring the length changes resulting from changes in the loading, or for that matter the length changes resulting from changing the activity of the nerves to the muscular region of the muscle spindle. One needs to break the feedback loop at some point and to measure the response of the component parts of the system in the absence of feedback operation.

The question arises as to the best way in which the response of the various component parts of a system can be specified. Obviously what one wants to be able to do is to be able to predict the response of any particular part of the system for any input i.e. we want to be able to specify a function which produces the output from a given input:

$$\text{output} = \text{'function'} \times \text{input}$$

The required function is known as the *transfer function* of the system being tested.

$$\text{transfer function} = \frac{\text{output}}{\text{input}}$$

For the simple feedback loop discussed above

$$\text{transfer function} = \frac{A}{1 + A\beta}$$

$$= 1/\beta \qquad \text{if } A \gg 1$$

Although mathematically any input can be used, the two standard methods of analysis of the transfer function use as an input either a step change in the input (a sudden change from one level to another), or a sinusoidal input. The first is known as *transient analysis*, and the second as *sinusoidal analysis*. In transient analysis the output is measured as a function of time after applying the step change in the input; in sinusoidal analysis the relative phase and amplitude of the output are measured as compared to those of the input over as wide a range of different frequencies of sinusoidal oscillation of the input as possible.

If the response to an input of one magnitude is scaled in proportion compared with the input of another magnitude (i.e. for an input n times greater the response is n times greater), and if the response to increase in the input is the inverse of the response to a decrease in the input, then the system is said to be *linear*; otherwise it is said to be *nonlinear*. The analysis of linear systems is fairly straightforward; that of nonlinear systems is a complex and horrendous undertaking!

Solutions to Questions

Chapter 1

1. (i) $\dfrac{d(a \cdot \sin(bt))}{dt} = ab \cdot \cos(bt)$

(ii) $\dfrac{d(ae^{bt})}{dt} = abe^{bt}$

(iii) $\dfrac{d(t \cdot \cos(bt))}{dt} = t \cdot \dfrac{d(\cos(bt))}{dt} + \cos(bt) \cdot \dfrac{dt}{dt}$

$$= -bt \cdot \sin(bt) + \cos(bt)$$

(iv) $\dfrac{d(e^{bt} \cdot \cos(at))}{dt} = e^{bt} \cdot \dfrac{d(\cos(at))}{dt} + \cos(at) \cdot \dfrac{d(e^{bt})}{dt}$

$$= e^{bt}(-a \cdot \sin(at)) + \cos(at) \cdot be^{bt}$$

$$= -ae^{bt}\sin(at) + be^{bt}\cos(at)$$

(v) $\dfrac{d(\cos(bt)/t)}{dt} = \dfrac{\cos(bt) \cdot \dfrac{dt}{dt} - t \cdot \dfrac{d(\cos(bt))}{dt}}{t^2}$

$$= \frac{1}{t^2}\ \{\cos(bt) - t \cdot (-b \cdot \sin(bt))\}$$

$$= \{\cos(bt) + bt \cdot \sin(bt)\}/t^2$$

(vi) Let $u = bt^2$

Then $\dfrac{d(a \cdot \sin(bt^2))}{dt} = \dfrac{d(a \cdot \sin(u))}{dt} = \dfrac{d(a \cdot \sin(u))}{du} \cdot \dfrac{du}{dt}$

$$= a \cdot \cos(u) \cdot \frac{d(bt^2)}{dt}$$

$$= a \cdot \cos(bt^2) \cdot 2bt = 2abt \cdot \cos(bt^2)$$

(vii) Let $v = bt^2$ and $u = e^{bt^2} = e^v$

Then $\dfrac{d(a \cdot \cos(e^{bt^2}))}{dt} = \dfrac{d(a \cdot \cos(u))}{dt} = \dfrac{d(a \cdot \cos(u))}{du} \cdot \dfrac{du}{dt}$

$$= -a \cdot \sin(u) \cdot \dfrac{d(e^{bt^2})}{dt}$$

$$= -a \cdot \sin(u) \cdot \dfrac{d(e^v)}{dt} = -a \cdot \sin(u) \cdot \dfrac{d(e^v)}{dv} \cdot \dfrac{dv}{dt}$$

$$= -a \cdot \sin(u) \cdot e^v \cdot 2bt = -2abt \cdot e^v \cdot \sin(u)$$

$$= -2abt \cdot e^{bt^2} \cdot \sin(e^{bt^2})$$

2. (a) Force $= \dfrac{d(\frac{1}{2}kx^2)}{dx} = \frac{1}{2}k \cdot 2x = kx$

   (b) Force $= \dfrac{d(e_1 \cdot e_2/x)}{dx} = -\dfrac{e_1 \cdot e_2}{x^2}$

3. $y = a \cdot \sin(bt)$

   Therefore, $\dfrac{dy}{dt} = ab \cdot \cos(bt)$

   and $\dfrac{d^2y}{dt^2} = -ab^2 \cdot \sin(bt) = -b^2 \cdot (a \cdot \sin(bt)) = -b^2y$

Both $y = c \cdot \cos(bt)$ and $y = a \cdot \sin(bt) + c \cdot \cos(bt)$ give

$$\dfrac{d^2y}{dt^2} = -b^2 \cdot y$$

Thus, if you meet the differential equation $\dfrac{d^2y}{dt^2} = -b^2y$ you know that all three of
the above expressions for y are possible solutions. The most general solution, because
it includes the other two, is

$$y = a \cdot \sin(bt) + c \cdot \cos(bt)$$

a and c are any constants. Their actual values cannot be determined unless more
information about the problem is provided.

4. If $y = Ae^{at}$, then $\dfrac{dy}{dt} = Aae^{at} = ay$.

   i.e. $\dfrac{dy}{dt} = ay$ has as one possible solution $y = Ae^{at}$.

(Note, there are other possible solutions).

In the population of animals, let n be the population, $\dfrac{dn}{dt_B}$ be the birth rate and $\dfrac{dn}{dt_D}$ be the death rate. Since both are proportional to the population we know that

$$\frac{dn}{dt_B} = B \cdot n \qquad \text{and} \qquad \frac{dn}{dt_D} = D \cdot n$$

in which B and D are the constants of proportionality. The net rate of change of the population is the difference between the birth and death rates, i.e.

$$\frac{dn}{dt} = \frac{dn}{dt_B} - \frac{dn}{dt_D} = B \cdot n - D \cdot n = (B - D) \cdot n$$

This resulting equation $\dfrac{dn}{dt} = (B - D) \cdot n$ is the same as $\dfrac{dy}{dt} = ay$ with n replacing y and $(B - D)$ replacing a. A solution to the equation for the population is thus

$$n = Ae^{(B-D) \cdot n}$$

5. (i) $\displaystyle\int a \cdot \sin(bt)\,dt = a \int \sin(bt) \cdot dt = -\frac{a}{b}\cos(bt) + C$

(ii) $\displaystyle\int -\frac{a}{y} \cdot dy = -a \int \frac{1}{y} \cdot dy = -a \cdot \ln(y) + C$

(iii) $\displaystyle\int b \cdot e^{at}\,dt = b \int e^{at}\,dt = \frac{b}{a} e^{at} + C$

(iv) $\displaystyle\int b \cdot e^{ay}\,dy = b \int e^{ay}\,dy = \frac{b}{a} e^{ay} + C$

(v) $\displaystyle\int_0^{\pi/2b} a \cdot \sin(bt) \cdot dt = \left[-\frac{a}{b}\cos(bt) \right]_0^{\pi/2b}$

$$= -\frac{a}{b}\left[\cos(\pi/2) - \cos(0)\right]$$

When discussing integration and differentiation of trigonometric functions (such as sin, cos, tan etc.) the angles must be specified in radians (see page 36). Now $\cos(\pi/2) = \cos(90°) = 0$ and $\cos(0) = 1$. Thus our answer is

$$-\frac{a}{b}(0 - 1) = a/b$$

(vi) $\displaystyle\int_0^{\pi/2b} a \cdot \sin(bt) \cdot dt = -\frac{a}{b}\left[\cos(\pi) - \cos(0)\right]$

$$= -\frac{a}{b}(-1 - 1) = 2a/b$$

(vii) $\int_0^1 x^2\,dx = \left[\frac{1}{3}x^3\right]_0^1 = \frac{1}{3}(1-0) = 1/3$

(viii) $\int_3^4 x^2\,dx = \left[\frac{1}{3}x^3\right]_3^4 = \frac{1}{3}(64-27) = 37/3$

Chapter 2

1. $60\,\dfrac{\text{mile}}{\text{hour}} \times \dfrac{1760\ \text{yard}}{\text{mile}} \times \dfrac{36\ \text{inch}}{\text{yard}} \times \dfrac{2\cdot54\ \text{cm}}{\text{inch}} \times \dfrac{\text{metre}}{100\ \text{cm}} \times \dfrac{\text{hour}}{3600\ \text{s}} = 26\cdot82\ \text{metre/s}$

2. Force $= k(\text{velocity})^x (\text{size})^y (\text{density})^z$

$mlt^{-2} = (lt^{-1})^x\, l^y\, (ml^{-3})^z$

Mass $1 = z$

Length $1 = x + y - 3z$

Time $-2 = -x$

   Thus $x = 2;\ y = 2;\ z = 1$

   Force $= k(\text{velocity})^2 (\text{size})^2 (\text{density})$

3. $1\,\dfrac{\text{gm cm}}{\text{s}^2} = \dfrac{1\ \text{gm cm}}{\text{s}^2}\ \dfrac{\text{kg}}{10^3\ \text{gm}}\ \dfrac{\text{m}}{10^2\ \text{cm}} = 10^{-5}\ \text{kg m/s}^2$

$= 10^{-5}\ \text{newton}$

4. Pressure $\cdot$ volume/gas constant $\cdot$ temperature

   Pressure $-$ force/area $= mlt^{-2}l^{-2}$

   Volume $- l^3$

   Gas const $-$ joule/$^\circ$K $= $ Nm K^{-1} $= mlt^{-2}l\ $K^{-1}

   Temperature $-$ K

$\dfrac{\text{Pressure} \cdot \text{volume}}{\text{Gas const} \cdot \text{temp}} - \dfrac{mlt^{-2}l^{-2}\ l^3}{mlt^{-2}lK^{-1}\,K} = \dfrac{ml^2\,t^{-2}}{ml^2\,t^{-2}} = 1$

i.e. PV/RT is dimensionless

5. $\dfrac{\text{Density length velocity}}{\text{viscosity}}$

$\dfrac{ml^{-3}\ l\ lt^{-1}}{ml^{-1}\,t^{-1}} = 1$ i.e. Reynold's number is dimensionless.

Chapter 3

1. This can be answered graphically or analytically. To find the solution graphically, draw a line 10 units long to represent the wind velocity, and through the top end of this line draw a line at 45° to represent the desired flight direction (see figure). With compasses set at 15 units draw an arc centred on the bottom end of the wind velocity line to cut the preferred direction line as indicated. The direction of flight of the locust is then given by the line AB, which is at about 73° to the direction of the wind.

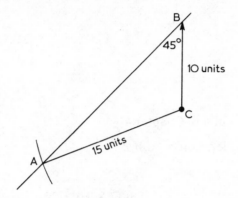

To solve the problem analytically requires the application of simple trigonometry to the above triangle. Use of the sine formula enables the angle BAC to be found directly:

$$\frac{10}{\sin BAC} = \frac{15}{\sin 45°}$$

i.e.

$$BAC = \sin^{-1}\left(\frac{10}{15 \times \sqrt{2}}\right) = \sin^{-1}(0{\cdot}47) = 28°$$

Thus direction to the wind $= 45° + 28° = 73°$

2.

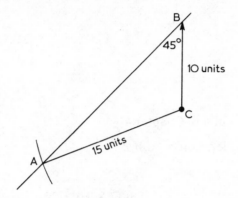

The action of the earth's gravity is vertical, that of the centrifugal force is horizontal. Thus the angle the test-tube makes with the vertical is

$$Angle = \tan^{-1}(\text{horizontal force/vertical force})$$

a. Angle $= \tan^{-1}(10/1) = 84{\cdot}3°$
b. Angle $= \tan^{-1}(100/1) = 89{\cdot}4°$
c. Angle $= \tan^{-1}(1000/1) = 89{\cdot}9°$

3. Take moments about the elbow (point E). Note that a mass of 2 kg exerts a force of $2 \times 9 \cdot 81 \doteq 20$ N due to the action of gravity, and that the mass of the arm can be considered to be concentrated at its centre of gravity in order to work out the moments due to the arm. The centre of gravity of a uniform rod is at its mid-point. Thus

$$2 \times 9 \cdot 81 \times 0 \cdot 3 + 0 \cdot 5 \times 9 \cdot 81 \times 0 \cdot 15 = \text{Tension} \times \sin(60°) \times 0 \cdot 03$$

$$\text{Tension} = \frac{9 \cdot 81 \, (0 \cdot 6 + 0 \cdot 075)}{0 \cdot 866 \times 0 \cdot 03} = 255 \, \text{N}$$

Chapter 4

1. $F = m \, GM/r^2$ equation 4.6, Chapter 4.

i.e. gravitational acceleration $= GM/r^2$

a. For the earth

$$\text{Gravitational acceleration} = \frac{6 \cdot 67 \times 10^{-11} \, \text{N} \, \text{m}^2 \, \text{kg}^{-2} \times 6 \times 10^{24} \, \text{kg}}{(6 \cdot 4 \times 10^6 \, \text{m})^2}$$

$$= \frac{6 \cdot 67 \times 6 \times 10^{13}}{6 \cdot 4 \times 6 \cdot 4 \times 10^{12}} \, \frac{\text{N} \, \text{m}^2 \, \text{kg}^{-1}}{\text{m}^2}$$

$$= 9 \cdot 77 \, \text{N} \, \text{kg}^{-1} = 9 \cdot 77 \, \text{kg} \, \text{m} \, \text{s}^{-2} \, \text{kg}^{-1}$$

$$= 9 \cdot 77 \, \text{m} \, \text{s}^{-2}$$

b. For the moon

$$\text{Gravitational acceleration} = \frac{6 \cdot 67 \times 10^{-11} \, \text{N} \, \text{m}^2 \, \text{kg}^{-2} \times 7 \cdot 4 \times 10^{22} \, \text{kg}}{(1 \cdot 74 \times 10^6 \, \text{m})^2}$$

$$= \frac{6 \cdot 67 \times 7 \cdot 4 \times 10^{11}}{1 \cdot 74^2 \times 10^{12}} \, \text{m} \, \text{s}^{-2}$$

$$= 1 \cdot 63 \, \text{m} \, \text{s}^{-2}$$

2. Force $= mr\omega^2$

$$= mr(2\pi f)^2$$

$$= 4\pi^2 r f^2 m$$

where f is the frequency of rotation of the earth $= 1/\text{day}$

Thus force $= 4\pi^2 m \times 6\cdot4 \times 10^6 m \times \dfrac{1}{\text{day}^2} \times \dfrac{\text{day}^2}{24^2\,\text{hour}^2} \times \dfrac{\text{hour}^2}{3600^2\,\text{s}}$

$\qquad = \dfrac{m \cdot 4\pi^2 \times 6\cdot4 \times 10^6}{24 \times 24 \times 3\cdot6 \times 3\cdot6 \times 10^6}\,\text{Kg m s}^{-2}$

$\qquad = 0\cdot043 \times m\,\text{Kg m s}^{-2}$

The acceleration due to the rotation of the earth is thus $0\cdot043$ m s^{-2} measured at the equator.

3. The normal beam balance works by comparing the gravitational force acting on a known mass, with that acting on an unknown mass. The readings will thus be the same at different latitudes.

The spring balance works by measuring the force acting on a mass via the extension of a spring. At the equator the centrifugal acceleration is $0\cdot043$ m s^{-2} (see question 2). At a latitude of $60°$ the centrifugal acceleration is $0\cdot043 \cos(60°)$ m s^{-2} = $0\cdot0215$ m s^{-2}. Thus the difference in weight as measured by the two balances will be $10 \times (0\cdot043 - 0\cdot0215) = 0\cdot215$ N.

4. The vertical fall of the cylinder is $5 \sin(30°) = 2\cdot5$ m. Thus if the cylinder has a mass of m kg, the loss of potential energy due to rolling down the plane is $2\cdot5\,m\,g$ joule, this energy will be converted into kinetic energy — both of translational and of rotational motion.

If the translational velocity is v m/s, and the rotational velocity is ω radians/s, then

$$\text{Kinetic energy} = \tfrac{1}{2}mv^2 + \tfrac{1}{2}I\omega^2$$

Since the radius is 2 cm = $0\cdot02$ m, then using equation 4.17 (Chapter 4)

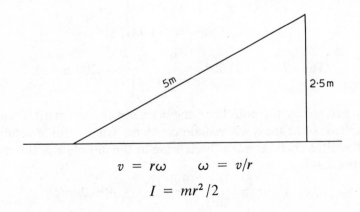

$$v = r\omega \qquad \omega = v/r$$

and table 5 $\qquad\qquad I = mr^2/2$

Thus, kinetic energy $= \frac{1}{2}mv^2 + \frac{1}{2}\left(\dfrac{mr^2}{2}\right)\dfrac{v^2}{r^2} = 0.75\,m\,v^2$

Thus,
$$2.5 \times m\,g = 0.75 \times m\,v^2$$

$$v = \left(\frac{2.5 \times 9.81}{0.75}\right)^{\frac{1}{2}} \text{m/s} = 5.72 \text{ m/s}$$

5. Since the orbit is steady, the gravitational attraction by the earth must equal the centrifugal force due to the circular motion of the spaceship;

$$m\,\frac{GM}{r^2} = m \cdot \frac{V^2}{r}$$

i.e
$$V^2 = GM/r = \frac{6.67 \times 10^{-11}\,\text{N}\,\text{m}^2\,\text{kg}^{-2} \times 6 \times 10^{24}\,\text{kg}}{(6400 + 2000) \times 10^3 \text{ m}^2}$$

$$= \frac{6.67 \times 6 \times 10^{13}}{8.4 \times 10^6}\,\frac{\text{kg m s}^{-2}\,\text{m}^2\,\text{kg}^{-2}\,\text{kg}}{\text{m}^2}$$

$$= 4.76 \times 10^7\,\text{m}^2\,\text{s}^{-2}$$

$$V = 6.9 \times 10^3 \text{ m/s}$$

6. If it is required to obtain the normal earth gravitational acceleration in the spaceship, then if the outer radius of the spaceship is R

$$\omega^2 R = 9.81 \text{ m/s}^2$$

The allowed tolerance of 1% change in acceleration at the astronaut's head necessitates that the acceleration at a radius $(R - 2)$ metres differ from that at R metres by only 1%

$$\text{i.e. } \omega^2(R) - \omega^2(R - 2) = 0.01 \times 9.81 = 0.0981$$

$$\omega^2(R - R + 2) = 0.0981$$

$$\omega^2 = 0.0981/2$$

$$\text{Thus } R = 9.81/\omega^2 = \frac{9.81 \times 2}{0.0981} = 200 \text{ metre}$$

7. If the force exerted by the muscle on the mass remains constant, then the acceleration (a) applied to the mass will remain constant. The net force acting on the mass will be the difference between the tension in the muscle and the gravitational attraction of the earth.

$$\text{i.e. } a = \frac{(5 - 0.1 \times 9.81)\,\text{N}}{0.1 \text{ kg}} = 40.19 \text{ m/s}^2$$

We can now use the equations of motion to obtain the position of the weight during the time that the force is being applied. $s = ut + \frac{1}{2}at^2 = \frac{1}{2}at^2$ (since the initial velocity is zero).

This will enable us to plot the position versus time. The position of the weight at the end of the period of muscle action is, for example, 0·2 metres above its starting value. The velocity of the mass at this time is given by

$$v = u + at = 0 + 40·19 \times 0·1 = 4·019 \text{ m/s}$$

Thus at the end of the period of muscle action, the mass is moving upwards at a velocity of 4·019 m/s. It is subsequently acted on solely by the acceleration due to gravity, and its motion is now given by

$$s = ut + 1/2at^2 = 4·019 - 9·81 \, t^2/2$$

The mass will hit the table after approximately a further 0·9 seconds.

8.

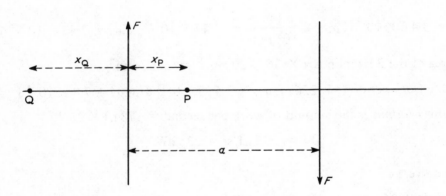

To establish that the moment of a couple about any point is constant we must consider the two cases (a) when the point lies between the opposing forces of the couple (b) when the point lies outside the line of action of the forces. These are illustrated in the above figure.

(a) Consider a point P between the line of action of the forces. Moment of the couple $= Fx_P + F(a - x_P) = Fa$
(b) Consider a point Q outside the lines of action. Moment of the couple $= -Fx_Q + F(a + x_Q) = Fa$

Note that we have taken clockwise moments as positive in all cases. Thus the moments are equal and independent of the position of the point chosen.

Chapter 5

1. In each case the Young's modulus is determined for the steepest part of the curves.

Bone $Y \doteq \text{stress/strain} \doteq \dfrac{300 \times 10^6 \, \text{N m}^{-2}}{0.01} = 3 \times 10^{10} \, \text{N m}^{-2}$

Hair $Y \doteq \dfrac{2.5 \times 10^6 \, \text{N m}^{-2}}{0.1} = 2.5 \times 10^7 \, \text{N m}^{-2}$

Resilin $Y \doteq \dfrac{1.4 \times 10^6 \, \text{N m}^{-2}}{1} = 1.4 \times 10^6 \, \text{N m}^{-2}$

2. 3 Hz curve

The length of the major axis is about 36 mm, and of the minor axis about 20 mm. The tension calibration is 10^{-4} N/18 mm, and the length calibration about 0·1 mm muscle/16 mm.

Thus 1 mm² on the paper corresponds to $\dfrac{10^{-4}}{18} \dfrac{0.1}{16}$ N mm

$$= 3.47 \times 10^{-8} \, \text{N mm} \times \frac{1 \, \text{m}}{10^3 \, \text{mm}} = 3.47 \times 10^{-11} \, \text{N m} = 3.47 \times 10^{-11} \, \text{J}$$

The area of the 3 Hz loop is $\pi \times 36 \times 20 \, \text{mm}^2 = 2262 \, \text{mm}^2$

Thus the *work per cycle* is $2262 \, \text{mm}^2 \times 3.47 \times 10^{-11} \, \text{J/mm}^2 \doteq 7.8 \times 10^{-8} \, \text{J} = 78 \, \text{nJ}$

The *power output* is the amount of work per second $= 78 \, \text{nJ} \times 3 \, \text{s}^{-1}$

$$= 235 \, \text{nJ/s} = 235 \, \text{nW}$$

25 Hz curve

Major axis $\doteq$ 36 mm;   Minor axis $\doteq$ 8·5 mm.   Area = 961 mm²

Work/cycle $= 961 \times 3.47 \times 10^{-11} \, \text{J} \doteq 33 \, \text{nJ}$

Power output $= 33 \, \text{nJ} \times 25 \, \text{s}^{-1} = 825 \, \text{nW}$

3. $Y = 10^{10} \, \text{N m}^{-2}$

Bone held vertically

Strain $= \text{stress}/Y = \text{force/unit area}/Y$

$$= \frac{1 \, \text{kg} \times 9.81 \, \text{m s}^{-2}}{0.005 \times 0.005 \, \text{m}^2 \times 10^{10} \, \text{N m}^{-2}} = \frac{9.81 \, \text{N}}{2.5 \times 10^5 \, \text{N}} = 3.9 \times 10^{-5}$$

Length change $= \text{strain} \times \text{overall length} = 3.9 \times 10^{-5} \times 100 \, \text{mm}$

$$= 3.9 \times 10^{-3} \, \text{mm} = 3.9 \, \mu\text{m}$$

Bone held horizontally

From page 52 we see that the deflection is $Pl^3/3YI_g$

For a square cross-section of side 5 mm I_g $=$ $(5 \times 10^{-3}\,\text{m})^4/12$

$$= 6{\cdot}25 \times 10^{-12}/12\,\text{m}^4$$

$$= 52{\cdot}1 \times 10^{-12}\,\text{m}^4$$

Thus deflection $= \dfrac{1\,\text{kg} \times 9{\cdot}81\,\text{m s}^{-2} \times 0{\cdot}1^3\,\text{m}^3}{3 \times 10^{10}\,\text{N m}^{-2} \times 52{\cdot}1 \times 10^{-12}\,\text{m}^4}$

$$= \dfrac{9{\cdot}81 \times 10^{-3}}{3 \times 52{\cdot}1 \times 10^{-2}}\ \dfrac{\text{kg m s}^{-2}\,\text{m}^3}{\text{kg m s}^{-2}\,\text{m}^{-2}\,\text{m}^4}$$

$$= 6{\cdot}27 \times 10^{-3}\,\text{m} = 6{\cdot}27\,\text{mm}$$

4. Since long-range elasticity is obtained from long chain molecules which are cross-linked at intervals sufficiently large that the intervening chains are highly flexible, whereas short-range elastic elements are usually ones with a high degree of rigid bonds; the implication is that the effect of strain is to break a number of the bonds which are present at low degrees of strain, thereby allowing the shapes of the ensuing chains to be randomised.

In fact, the protein keratin, which is the protein of hair, is α-helical at low degrees of strain. The α-helix is a helical winding of the protein chain, in which the turns of the helix are connected by the relatively weak hydrogen bonds. The effect of strain is to break the hydrogen bonds, whereupon the protein chain takes up what is known as the random-coil formation.

5. From page 58 we see that the work done in stretching an elastic link is $\tfrac{1}{2}kx^2$, where k is the stiffness, and x the extension.

Thus, $\qquad \tfrac{1}{2}k(10\,\text{nm})^2 = \dfrac{40}{2}\ \dfrac{\text{kJ}}{\text{mole}} \times \dfrac{\text{mole}}{6 \times 10^{23}\ \text{molecules}} \times \dfrac{10^3\ \text{J}}{\text{kJ}}$

$$k = \dfrac{40 \times 10^3}{6 \times 10^{23} \times 10^{-16}}\ \dfrac{\text{J}}{\text{m}^2\ \text{molecule}}$$

$$= 6{\cdot}7 \times 10^{-7} \times 10^3 = 6{\cdot}7 \times 10^{-4}\ \text{N m/m}^2$$

$$= 6{\cdot}7 \times 10^{-4}\ \text{N/m}$$

6. (a) In series. If the combines stiffness is k

$$\frac{1}{k} = \frac{1}{k_1} + \frac{1}{k_2}$$

(b) In parallel.

$$k = k_1 + k_2$$

7. From page 51, the flexural rigidity $= YI_g$

$$\text{strength to bending} = \text{const.}\, I_g/R_{max}$$

$$\text{and} \qquad\qquad I_g = A(R^2 - A/2\pi)/2$$

For two skeletons of the same cross-sectional area, but of external radius R_1 and $10R_1$

$$I_{g1} = \frac{AR^2}{2} - \frac{A^2}{4\pi}$$

$$I_{g10} = \frac{100AR^2}{2} - \frac{A^2}{4\pi}$$

For the skeletons of external radius 1 mm

$$I_{g1} = \tfrac{3}{2}(1 - 3/2\pi)\,\text{mm}^4 = 0{\cdot}78\,\text{mm}^4$$

For that of radius 5 mm

$$I_{g5} = \tfrac{3}{2}(25 - 3/2\pi) = 36{\cdot}8\,\text{mm}^4$$

Thus the *ratio of the flexural rigidity* in these two skeletons is $36{\cdot}8/0{\cdot}78 \doteq 47$

The *ratio of the strength to bending* is $\dfrac{36{\cdot}8/5}{0{\cdot}78/1} = 9{\cdot}4$

Chapter 6

1. Blood flow $= 5\,\text{l/m}$, through tube 1 cm in radius.

Cross sectional area $= \pi \times 0{\cdot}01^2\,\text{m}^2 = 3{\cdot}14 \times 10^{-4}\,\text{m}^2$.

Volume of blood $= 5\text{l} = 5 \times 10^3\,\dfrac{\text{cm}^3 \times \text{m}^3}{10^6\,\text{cm}^3} = 5 \times 10^{-3}\,\text{m}^3$

Blood velocity $=$ volume flow/unit time/cross-sectional area

$$= \frac{5 \times 10^{-3}\,\text{m}^3}{\text{min}} \times \frac{\text{min}}{60\,\text{s}} \times \frac{1}{3{\cdot}14 \times 10^{-4}\,\text{m}^2}$$

$$= \frac{5 \times 10^{-3}}{60 \times 3{\cdot}14 \times 10^{-4}}\,\text{m/s}$$

$$= 0{\cdot}265\,\text{m/s}$$

Critical velocity (equation 6.3, page 71) $= 100 \times \eta/\rho \times a$

$$= \frac{1000 \times 4 \times 10^{-3}\,\text{N s m}^{-2}}{10^3\,\text{kg m}^{-3} \times 0{\cdot}01\,\text{m}}$$

$$= 0{\cdot}4\,\frac{\text{kg m s}^{-2}\,\text{s m}^{-2}}{\text{kg m}^{-3}\,\text{m}} = 0{\cdot}4\,\text{m s}^{-1}$$

To reach the critical velocity the pumping rate must be increased in the ratio $0.4/0.265 = 1.51$. That is, the blood flow must exceed $5 \times 1.51 = 7.55$ l/m

2. Cross-sectional area of a capillary $= \pi \times (8 \times 10^{-6})^2 \, m^2$

$$= 2.01 \times 10^{-10} \, m^2$$

Cross-sectional area of 25×10^6 capillaries $= 25 \times 10^6 \times 2.01 \times 10^{-10}$

$$= 50.3 \times 10^{-4} \, m^2$$

Thus, for a pumping rate of 5 l/min,

$$\text{Average blood velocity} = \frac{5 \times 10^{-3} \, m^3}{60 \, s} \times \frac{1}{50.3 \times 10^{-4} \, m^2}$$

$$= 0.0166 \, m/s$$

From equation 6.5, volume/s $= \pi P a^4 / 8 L \eta$

i.e. pressure difference $= 8 L \eta (\text{volume/s})/\pi a^4$

$$= \frac{8 \times 10^{-3} \, m \times 5 \times 10^{-3} \, m^3 \times 4 \times 10^{-3} \, N \, s \, m^{-2}}{60 \, s \times \pi \times (4 \times 10^{-6} \, m)^4 \times 25 \times 10^6}$$

$$= \frac{8 \times 5 \times 4 \times 10^{-9}}{60 \times \pi \times 256 \times 25 \times 10^{-18}} \quad \frac{m \, m^3 \, N \, s \, m^{-2}}{s \, m^4}$$

$$= 1.33 \times 10^5 \, N/m^2$$

Note 1 atmosphere $= 1.01 \times 10^5 \, N/m^2$.

3. Reynold's number $=$ density $\times$ velocity $\times$ length/viscosity.

For air, density $= 1.29 \, kg/m^3$

viscosity $= 1.82 \times 10^{-5} \, N \, s \, m^{-2}$

Assume that a locust can glide at $2 \, m \, s^{-1}$, and that its abdomen is 5 cm long, then

$$R_n \text{ (locust)} = \frac{1.29 \times 2 \times 0.05}{1.82 \times 10^{-5}} \quad \frac{kg \, m^{-3} \, m \, s^{-1} \, m}{kg \, m \, s^{-2} \, s \, m^{-2}}$$

$$\doteq 7000$$

Assume a midge can fly at 10 cm/s, and has a length of 1 mm.

Then, $$R_n \text{(midge)} = \frac{1.29 \times 0.1 \times 10^{-3}}{1.82 \times 10^{-5}} \doteq 7$$

4. (i) From Stoke's Law (equation 6.7 page 75) the force on a molecule moving through a fluid is

$$\text{Force} = 6\pi v \eta a$$

$$\text{Viscosity of water} = 10^{-3}\ \text{N s m}^{-2}$$

$$\text{Force} = 6\pi \times 10^{-6}\ \text{m s}^{-1} \times 10^{-3}\ \text{N s m}^{-2} \times 2{\cdot}5 \times 10^{-9}\ \text{m}$$

$$= 4{\cdot}7 \times 10^{-17}\ \text{N}$$

The gravitational force acting on the same molecule $= \text{mass} \times g$

$$= \frac{5 \times 10^4\ g}{6 \times 10^{23}} \times 9{\cdot}81\ \text{m s}^{-2} \times 10^{-3}\ \frac{\text{kg}}{g}$$

$$= 8{\cdot}2 \times 10^{-24}\ \text{N}$$

(ii) The centrifugal force acting on a molecule in a fluid is given by

$$\text{Force} = \text{mass} \times \text{acceleration}$$

However, the mass relevant to this calculation is not the mass of the molecule, but the mass of the molecule minus the mass of the fluid displaced (Archimedes principle).

$$\text{Mass of molecule of M. Wt 43 000 daltons} = \frac{43\ \text{kg}}{6 \times 10^{23}} = 7{\cdot}17 \times 10^{-23}\ \text{kg}$$

$$\text{Mass of water displaced} = \tfrac{4}{3}\pi \times (2{\cdot}5 \times 10^{-9})^3 \times 10^3\ \text{kg}$$

$$= 6{\cdot}54 \times 10^{-23}\ \text{kg}$$

$$\text{Effective mass} = (7{\cdot}17 - 6{\cdot}54) \times 10^{-23} = 0{\cdot}63 \times 10^{-23}\ \text{kg}$$

Therefore, required acceleration

$$= \frac{4{\cdot}7 \times 10^{-17}\ \text{N}}{0{\cdot}63 \times 10^{-23}\ \text{kg}} = 7{\cdot}5 \times 10^6\ \text{m s}^{-2}$$

$$\text{Acceleration in circular motion} = r\omega^2 = 4\pi^2 f^2 r$$

Thus:
$$f^2 = \frac{7{\cdot}5 \times 10^6\ \text{m s}^{-2}}{4 \times \pi^2 \times 0{\cdot}1\ \text{m}} = 1{\cdot}9 \times 10^6\ \text{s}^{-2}$$

$$f = 1{\cdot}37 \times 10^3\ \text{revolutions per second}$$

Chapter 7

1. From page 93 we see that the capillary rise h is given by $h = \dfrac{2\,T\cos\theta}{\rho g r}$

For water, $T = 7{\cdot}3 \times 10^{-2}\ \text{N} \cdot \text{m}^{-1}$, $\rho = 10^3\ \text{kg/m}^3$
Thus, for $\theta = 0°$ $(\cos\theta = 1)$

$$h = \frac{2 \times 7 \cdot 3 \times 10^{-2} \text{ N} \cdot \text{m}^{-1} \times 1}{10^3 \text{ kg} \cdot \text{m}^{-3} \times 9 \cdot 81 \text{ m} \cdot \text{s}^{-2} \times 25 \times 10^{-6} \text{ m}}$$

$$= \frac{2 \times 7 \cdot 3 \times 10^{-2}}{9 \cdot 81 \times 25 \times 10^{-6} \times 10^3} \quad \frac{\text{N m}^3 \text{ s}^2}{\text{m kg m m}}$$

$$= 0 \cdot 0595 \times 10 \frac{\text{kg} \cdot \text{m}}{\text{s}^2} \times \frac{\text{s}^2}{\text{kg}}$$

$$= 0 \cdot 595 \text{ m} = 595 \text{ mm}$$

For $\theta = 60°$ ($\cos \theta = 0 \cdot 5$)

$h = 595 \times 0 \cdot 5 = 297 \cdot 5$ mm

2. The water will rise 595 mm up the column as described in question 1.
If the tube is only protruding 10 mm above the surface the water will not flow over the top — this would violate the principle of conservation of energy and be a perpetual motion machine. The reason that it will not overflow is that at the top of the tube the edge of the meniscus of the water will no longer be held vertically since the glass in contact with the water is no longer vertical. The radius of curvature of the water will thus become much greater, thereby reducing the pressure difference across the surface.

3. Length of water surface in contact with the stick is equal to the circumference of the stick $(2\pi r) = 2 \times 3 \cdot 1416 \times 2$ mm $= 12 \cdot 57$ mm $= 0 \cdot 01257$ m
Force acting on stick $= T \cdot 2\pi r = 7 \cdot 3 \times 10^{-2}$ N $\cdot$ m^{-1} $\times 0 \cdot 01257$ m $= 9 \cdot 17 \times 10^{-4}$ N
Work $=$ Force $\times$ distance moved $=$ force $\times$ length of stick $= 9 \cdot 17 \times 10^{-4}$ N $\times 0.02$ m
$= 18 \cdot 3 \times 10^{-6}$ N $\cdot$ m $= 18 \cdot 3 \ \mu$J.

Chapter 8

1.
$$K = \frac{\text{ATP AMP}}{\text{ADP}^2} \qquad \text{AMP} = \frac{K \text{ ADP}^2}{\text{ATP}}$$

i.e. $\text{AMP} = \dfrac{2 \cdot 27 \times (5 \times 10^{-4})^2}{8 \times 10^{-3}} = \dfrac{2 \cdot 27 \times 25 \times 10^{-8}}{8 \times 10^{-3}}$ M

$= 7 \cdot 1 \times 10^{-5}$ M $= 71 \ \mu$M

Free energy change $n \, \Delta G = n\Delta G_0 + nRT \ln \dfrac{\text{ATP AMP}}{\text{ADP}^2}$

This is the free energy change to convert one mole of each of the reactants into one mole of each of the products, or to convert one mole of each of the products into one mole of each of the reactants. In this example, the product of the reaction is ADP, but two molecules of ADP are formed by the interaction of ATP and AMP. Thus, if $n = 1$ in the above equation, this gives the free energy change for the conversion of two mole of ADP.

However, since the reaction is at equilibrium, there is no change in the free energy as the reaction proceeds, provided that the concentrations are unchanged due to the progress of the reaction (this of course requires back-up systems to control the concentrations).

2. Work is done in transferring a sodium ion, due both to the concentration gradient and to the electrical gradient.

Work against chemical gradient $= \mu_0 + RT \ln(10) - \mu_0 - RT \ln(250)$

$$= RT \ln(10/250) = 8 \cdot 32 \, \text{J K}^{-1} \, \text{mole}^{-1} \times 300 \, \text{K} \times (-3 \cdot 22)$$

$$= -8030 \, \text{J/mole}$$

Work against electrical gradient/mole $=$ charge/mole $\times$ voltage change

$$= 96\,500 \, \text{C mole}^{-1} \times (-0 \cdot 06) \, \text{V} = 5790 \, \text{J}$$

Thus total work done in crossing membrane $= -8030 - 5790 \, \text{J/mole}$

$$= -13 \cdot 8 \, \text{kJ/mole}$$

The negative sign signifies that work must be expended to cause the ion to cross the membrane.

Free energy change obtained from the hydrolysis of ATP under the conditions specified

$$\Delta G = -7000 \, \frac{\text{cal}}{\text{mole}} \times 4 \cdot 19 \, \frac{\text{J}}{\text{cal}} + RT \ln \frac{5 \times 10^{-4} \times 2 \times 10^{-2}}{8 \times 10^{-3}}$$

$$= -29\,330 + 8 \cdot 32 \times 300 \times \ln(1 \cdot 25 \times 10^{-3}) \, \text{J}$$

$$= -29\,330 + 2500 \times (-6 \cdot 68) \, \text{J}$$

$$= -46\,040 \, \text{J}$$

Thus the hydrolysis of one molecule of ATP has a free energy change about three times that required to transport one Na^+ across the membrane under conditions approximately those found in nerve.

3. Work in expanding a gas from a pressure p_1 to a pressure p_2 is

$$\int_{p_1}^{p_2} p \cdot dV$$

Now, for a perfect gas, $pV = mRT$ where m is the number of gram moles of gas present. i.e.

$$V = \frac{mRT}{p}$$

$$\frac{dV}{dp} = -\frac{mRT}{p^2}$$

$$dV = -\frac{mRT}{p^2} \cdot dp$$

Thus, work in expansion $= \int_{p_1}^{p_2} p \cdot dV = \int_{p_1}^{p_2} p \cdot \frac{(-mRT)}{p^2} \cdot dp$

$$= \int_{p_1}^{p_2} -\frac{mRT}{p} \, dp = -mRT(\ln(p_2) - \ln(p_1))$$

$$= mRT(\ln(p_1) - \ln(p_2)) = mRT \ln\frac{p_1}{p_2}$$

4. For a 1% solution of KCl in water:

i. Density $= 1 \cdot 0046 \times 10^3$ kg solution/m^3.

ii. Volume/kg solution $= 1/\text{density} = 1/(1 \cdot 0046 \times 10^3)$ m^3/kg solution

$$= 0 \cdot 9954 \times 10^3 \text{ m}^3/\text{kg solution}$$

iii. Volume/kg solvent (water) $= \dfrac{0 \cdot 9954 \times 10^{-3}}{0 \cdot 99}$ m^3/kg water

$$= 1 \cdot 005476 \times 10^{-3} \text{ m}^3/\text{kg water}$$

(We divided by $0 \cdot 99$ since the solution is $0 \cdot 99$ water).

iv. Mass KCl/kg water $= \dfrac{0 \cdot 01}{0 \cdot 99}$ kg $= 0 \cdot 0101$ kg

In the same manner we can determine the volume per kg water ($1 \cdot 009306 \times 10^{-3}$ m^3/kg water) and mass of KCl per kg water ($0 \cdot 0204$ kg) for a 2% solution.

Thus to convert a solution containing 1 kg water from a 1% to 2% solution we have to add $0 \cdot 0204 - 0 \cdot 0101 = 0 \cdot 0103$ kg; the resulting volume change is then ($1 \cdot 009306 - 1 \cdot 005476$) $\times 10^{-3}$ m^3. Thus 1 g-mole (74 g) of KCl would (by proportionality) cause a volume change of

$$\frac{(1 \cdot 009306 - 1 \cdot 005476) \times 10^{-3} \text{ m}^3}{0 \cdot 0103} \times 74 \text{ g} \times \frac{\text{kg}}{10^3 \text{ g}}$$

$$= \frac{0 \cdot 00383 \times 74 \times 10^{-6}}{0 \cdot 0103} \text{ m}^3 = 27 \cdot 5 \times 10^{-6} \text{ m}^3 = 27 \cdot 5 \text{ cm}^3$$

The calculations for the partial molar volume of a 21% solution proceed in exactly the same way. Thus the volume/kg water of a 20% solution is

$$\frac{1}{1 \cdot 1328 \times 10^3 \times 0 \cdot 80} = 1 \cdot 10346 \times 10^{-3} \text{ m}^3,$$

and of a 22% solution is

$$\frac{1}{1 \cdot 1474 \times 10^3 \times 0 \cdot 78} = 1 \cdot 11735 \times 10^{-3} \text{ m}^3,$$

the difference in volume being 13·89 cm^3. The mass of KCl per kg water of a 20% solution is 0·2/0·8 = 0·25 kg, and of a 22% solution is 0·22/0·78 = 0·28205 kg, the difference being 32·05 g. Therefore the partial molar volume of a 21% solution of KCl is 13·89 × 74/32·05 = 32·07 cm^3 (closer to the volume of 1 g-mole of crystalline KCl than the partial molar volume of a 1% solution).

5. The calculation of the partial molar volume of water proceeds on the same lines as that for the partial molar volume of KCl in question 4. However, instead of determining the volume of the solution for a standard amount of water, we determine the volume for a standard amount of KCl; We also determine the mass of water in the solution for the same standard amount of KCl.
  For a 20% solution:

  i. Density = 1·1328 × 10^3 kg solution/m^3

  ii. Volume/kg solution = 1/density = 0·88277 × 10^{-3} m^3/kg solution

  iii. Volume/kg KCl = $\dfrac{0 \cdot 88277 \times 10^{-3}}{0 \cdot 2}$ = 4·41384 × 10^{-3} m^3/kg KCl

  iv. Mass water/kg KCl = 0·8/0·2 = 4 kg water/kg KCl.

Likewise for the 22% solution we find that the volume/kg KCl is

$$\frac{1}{1 \cdot 1474 \times 10^3 \times 0 \cdot 22} = 3 \cdot 96153 \times 10^{-3} \text{ m}^3/\text{kg KCl.}$$

The difference in volumes is thus (4·41384 − 3·96153) × 10^{-3} m^3 = 0·4523 × 10^{-3} m^3. The mass of water/kg KCl is 0·78/0·22 = 3·5454 kg, the difference in mass of water between the two solutions being 4 − 3·54545 = 0·4546 kg. Thus the partial molar volume of water is

$$\frac{0 \cdot 4523 \times 10^{-3} \text{ m}^3}{0 \cdot 4546 \text{ kg}} \times 18 \text{ g} \times \frac{10^{-3} \text{ kg}}{\text{g}} = 17 \cdot 91 \text{ cm}^3.$$

This is rather less than the molar volume of pure water (about 18 cm^3).

6. If sea water contains 500 mM NaCl, then the concentration is 0·5 moles/litre. Osmotic pressure $\pi = RTc$

$$= 8\cdot32 \frac{J}{K \cdot mole} \times 300\ K \times 0\cdot5 \frac{mole}{litre}$$

$$= 1248\ J/litre$$

$$= 1248 \frac{N\,m}{litre} \times \frac{10^3\ litre}{m^3} = 1\cdot248 \times 10^6\ N\,m^{-2}$$

1 atmosphere $= 1\cdot01 \times 10^5\ N\,m^{-2}$. Thus, measured in atmospheres the osmotic pressure is $1\cdot248 \times 10^6 / 1\cdot01 \times 10^5 = 12\cdot36$ atmospheres.

Chapter 9

1. (a) Energy difference between states A and B (caused by the extension of 10 nm of the elastic link of stiffness 4×10^{-4} N/m) (using equation 5.19 on page 57).

$$Energy\ difference = 1/2kX^2 = 0\cdot5 \times 4 \times 10^{-4}\,N\,m^{-1} \times (10^{-8}\,m)^2$$

$$= 2 \times 10^{-20}\,Nm = 2 \times 10^{-20}\,J$$

Using the Boltzmann distribution (equation 9.9b)

Ratio of cross bridges in the two states

$$\frac{No.\ in\ state\ B}{No.\ in\ state\ A} = exp\left(\frac{-2 \times 10^{-20}\,J}{1\cdot38 \times 10^{-23}\,J\,K^{-1} \times 300\,K}\right)$$

$$= exp(-4\cdot83) = 7\cdot98 \times 10^{-3} = 1/125$$

(b) When the elastic link is stretched by 5 nm in state A (and thus by 15 nm in state B) the energy difference between the two states is:

$$\tfrac{1}{2}k\ (X_2^2 - X_1^2) = 0\cdot5 \times 4 \times 10^{-4} \times (225 - 25) \times 10^{-18}\,J$$

$$= 4 \times 10^{-20}\,J$$

Thus, $$\frac{No.\ in\ state\ B}{No.\ in\ state\ A} = exp\left(\frac{-4 \times 10^{-20}}{0\cdot414 \times 10^{-20}}\right) = exp(-9\cdot66)$$

$$= 6\cdot4 \times 10^{-5} = 1/15\ 700$$

For case (a), difference in energy between states

$$= 2 \times 10^{-20}\ J/molecule$$

$$= \frac{2 \times 10^{-20}}{6 \times 10^{-23}}\ J/mole = 333\ J/mole$$

2. Surface area (neglecting the ends) $= \pi \times 0.3 \times 2$

$$= 1.88 \text{ m}^2$$

From equation 9.14, heat loss/second $= e\sigma(T_1^4 - T_2^4)$ area

Net heat loss/s $= 0.9 \times 5.69 \times 10^{-8} \text{ W m}^{-2} \text{ K}^{-4} \times (298^4 - 278^4) \text{ K}^4 \times 1.88 \text{ m}^2$

$$= 0.9 \times 5.69 \times 1.88 \times (1.91 \times 10^9) \times 10^{-8} \text{ W}$$

$$= 184.4 \text{ W}$$

The heat loss from the surface of the animal is $e\sigma A(T_1^4 - T_2^4)$ where A is the surface area, and T_1 is the surface temperature and T_2 the surrounding temperature. Provided that the thickness of fur required is sufficiently low, then the surface area of the animal will not change significantly, and the temperature of the outer surface of the fur necessary to cause a 2-fold decrease in heat loss can be determined.

$$\frac{\text{heat loss}}{0.5 \times \text{heat loss}} = \frac{e\sigma A(298^4 - 278^4)}{e\sigma A(T_s^4 - 278^4)}$$

$$T_s^4 - 278^4 = 0.5 \times (298^4 - 278^4) = 0.5 \times 1.91 \times 10^9$$

$$= 9.57 \times 10^8$$

$$T_s^4 = 278^4 + 9.57 \times 10^8 = 5.972 \times 10^9 + 9.57 \times 10^8$$

$$= 6.929 \times 10^9$$

$$T_s = 288.5°\text{K}$$

We know the rate of heat loss from the animal (92.2 W) and the temperature on either side of the fur (298°K and 288.5°K). Assuming once again that the thickness of fur is small compared to the radius of the animal, we can determine the required thickness of fur on the basis that the fur is a flat sheet, using equation 9.18a.

$$Q/t = kA(T_1 - T_2)/\text{thickness}$$

i.e. thickness required $= kA(T_1 - T_2)/\text{rate of heat loss}$

$$= \frac{0.035 \text{ J m}^{-1} \text{ s}^{-1} °\text{C}^{-1} \times 1.88 \text{ m}^2 \times 9.5°\text{C}}{92.2 \text{ J s}^{-1}}$$

$$= 0.0067 \text{ m}$$

$$= 6.7 \text{ mm}$$

3.

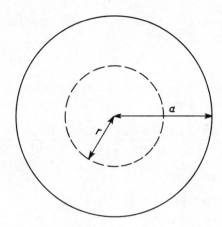

The question differs from example 9.1 in the geometry of the object being considered. In the example the situation was diffusion into or out of a cylinder, here it concerns diffusion into a sphere. Using the same terminology as for example 9.1, we now have:

$$\frac{dn}{dt} = D\,4\pi r^2\,\frac{dc}{dr} = \frac{4}{3}\pi r^3 B$$

$$\frac{dc}{dr} = \frac{Br}{3D}$$

giving

$$c(r) = C_0 - \frac{B}{6D}(a^2 - r^2)$$

Note that the units of B are mole $m^{-3}\,s^{-1}$

units of $c(r)$ and C_0 are mole m^{-3}

units of D are $m^2\,s^{-1}$

Now the rate of oxygen consumption is given as 1 ml/gm wet wt./hour

$$B = \frac{1\ ml}{g\ hour} \times \frac{10^{-6}\ m^3}{ml} \times \frac{10^3\ moles\ O_2}{22.4\ m^3} \times \frac{10^3\ g}{kg} \times \frac{hour}{3600\ s} \times \frac{10^3\ kg}{m^3}$$

$$= \frac{1}{22.4 \times 3600} \times 10^3\ mole\ m^{-3}\ s^{-1}$$

$$= 0.0124\ mole\ m^{-3}\ s^{-1}$$

The partial pressure of the oxygen in the water surrounding the animal is atmospheric. The concentration of oxygen in the water is thus given by the solubility product α times the concentration of oxygen in the atmosphere (n.b. oxygen forms 0.2 of normal air)

$$C_0 = \frac{0.2 \times 0.03 \times 10^3\ mole\ O_2}{22.4\ m^3} = 0.268\ mole/m^3$$

The partial pressure of oxygen at the centre of the animal is 0·01 atmospheres, which gives a concentration 1/20 that of C_0

$$\text{i.e. } C(\text{centre}) = 0.25/20 = 0.013 \text{ mole/m}^3$$

Thus, using the above equation to find the maximal radius of the animal, a (with r in the above equation equal to zero)

$$a^2 = \frac{(0.268 - 0.013) \text{ mole m}^{-3} \times 6 \times 1.72 \times 10^{-9} \text{ m}^2 \text{ s}^{-1}}{0.0124 \text{ mole m}^{-3} \text{ s}^{-1}}$$

$$= 2.12 \times 10^{-7} \text{ m}^2$$

$$\text{i.e. } a = 4.61 \times 10^{-4} \text{ m} = 0.461 \text{ mm}$$

4. Oxygen consumption of a single animal:

$$\frac{0.5 \text{ ml}}{\text{g wt hour}} \times \frac{1 \text{ gm wt}}{9.81 \text{ N}} \times 0.01 \text{ N} \times \frac{1 \text{ hour}}{3600 \text{ s}} \times \frac{10^{-3} \text{ mole}}{22.4 \text{ ml}} \text{ O}_2$$

$$= 6.32 \times 10^{-12} \text{ mole/s}$$

From question 3, concentration of O_2 in water at normal atmospheric pressure of air is 0·268 mole/m^3.

Thus, using equation 9.19

$$\frac{dn}{dt} = DA \, dc/dx = 1.72 \times 10^{-9} \text{ m}^2 \text{ s}^{-1} \times 1 \text{ m}^2 \times \frac{0.268 \text{ mole m}^{-3}}{10 \text{ m}}$$

$$= 4.61 \times 10^{-11} \text{ mole/s.}$$

The surviving density of animals (assuming that they can exist at fractionally above zero oxygen concentration) is thus

$$\frac{4.61 \times 10^{-11}}{6.32 \times 10^{-12}} = 7.3 \text{ animals/m}^2$$

5. This question requires that you estimate the volume of your exhaled breath, and also the rate of your breathing. You could measure the volume in a crude way by breathing through a tube into a measuring cylinder of water inverted in the sink. The volume will be approximately 1 litre. A normal breathing rate is about 20/minute, or 0·33 s^{-1}. Thus, the average rate of expired air loss is 0·33 l/s = 3.3×10^{-4} m^3/s. If 7% of this is water vapour, the rate of loss of water is about $3.3 \times 10^{-4} \times 0.07 = 2.3 \times 10^{-5}$ m^3/s at atmospheric pressure.

Thus, rate of water loss $= 2\cdot3 \times 10^{-5}\ \mathrm{m^3\ s^{-1}} \times \dfrac{10^3\ \mathrm{mole}}{22\cdot4\ \mathrm{m^3}}$

$$= 1\cdot03 \times 10^{-3}\ \mathrm{mole/s}$$

$$= 1\cdot03 \times 10^{-3}\ \mathrm{mole\ s^{-1}} \times 0\cdot018\ \mathrm{kg/mole}$$

$$= 1\cdot85 \times 10^{-5}\ \mathrm{kg\,s^{-1}}$$

The latent heat of vaporization of water is $2\cdot256 \times 10^6\ \mathrm{J\,kg^{-1}}$. Thus, rate of loss of energy

$$= 1\cdot85 \times 10^{-5} \times 2\cdot256 \times 10^6\ \mathrm{J\,s^{-1}} = 41\cdot9\ \mathrm{W}$$

Chapter 10

1. Equation 10.26 gives the wavelength of electrons accelerated through a voltage V as $1\cdot23/\!\sqrt{V}$ nm. The wavelength with an acceleration voltage of 80 kV is thus $1\cdot23/\!\sqrt{(8\cdot10^4)} = 0\cdot0043$ nm.

The resolving power of a microscope (equation 13.21) is $0\cdot61\lambda/$(numerical aperture). This applies as much to the electron microscope as to the light microscope. The resolving power is thus $0\cdot61 \times 0\cdot0043/0\cdot005 = 0\cdot53$ nm.

Chapter 11

1. The thermal noise of a resistor is $\sqrt{(4kTBR)}$. At room temperature (say 300 K) the RMS noise in a 1000Ω resistor at a bandwidth of 100 kHz is $(4 \times 1\cdot38 \times 10^{-23} \times 300 \times 10^5 \times 10^3)^{\frac{1}{2}}\ V = 1\cdot29\ \mu V$.

Similarly the RMS noise in a 1 MΩ resistor is $40\cdot7\ \mu V$.

2. In Fig. 11.10, if the electrodes have impedances of 20 MΩ each, and the input impedance of the amplifier is 2 MΩ then (from equation 11.1) the voltage across the input impedance will be $2/(2 \times 20 + 2) = 2/42 = 0\cdot048$ of the voltage across the membrane.

Chapter 15

1. One electron-volt is the work done in transporting an electron (of charge $1\cdot6 \times 10^{-19}$ coulomb) through 1 volt. Now 1 joule equals 1 coulomb volt. Thus the electron-volt $= 1\cdot6 \times 10^{-19}$ J.

2. From equation 15.5, $t_{\frac{1}{2}} = 0.693/\lambda$

Thus, $$\lambda_{Na} = \frac{0.693}{15 \text{ hour}} \times \frac{1 \text{ hour}}{3600 \text{ s}} = 1.28 \times 10^{-5} \times \text{s}^{-1}$$

$$N_t = N_0 \exp(-\lambda t)$$

$$\lambda t = \log_e (N_0/N_t)$$

Thus, time taken to reach 1/10 activity $= \dfrac{\log_e (10) \text{ s}}{1.28 \times 10^{-5}}$

$$= 1.8 \times 10^5 \text{ s}$$

$$= 50 \text{ hours}$$

3. $$12\,000 = 42\,000 \exp(-\lambda \times 1 \text{ day})$$

$$\lambda = \frac{\log_e (42/12)}{1 \text{ day}} = 1.25 \text{ day}^{-1}$$

$$t_{\frac{1}{2}} = 0.693/1.25 = 0.55 \text{ day}$$

4. No. of molecules of ATP $= 6 \times 10^{23}/551 = 1.09 \times 10^{21}$

No. of disintegrations/s $= 3.7 \times 10^{10} \times 10^{-4} = 3.7 \times 10^6$ dps.

Half-life of $P^{32} = 14.2$ days

$$\therefore \lambda = \frac{0.693}{14.2 \text{ d}} \times \frac{1 \text{ d}}{24 \text{ h}} \times \frac{1 \text{ h}}{3600 \text{ s}} = 5.65 \times 10^{-7} \text{ s}^{-1}$$

$$\frac{dN}{dt} = -\lambda N \qquad \text{i.e. } N = \frac{1}{\lambda} \frac{dN}{dt}$$

i.e. $$N = \frac{3.7 \times 10^6}{5.65 \times 10^{-7}} = 6.55 \times 10^{12}$$

Therefore, the fraction of the ATP which have P^{32} is

$$\frac{6.55 \times 10^{12}}{1.09 \times 10^{21}} = 6 \times 10^{-9}$$

5. The formula giving the absorption of rays is equation 15.6

$$I = I_0 e^{-\mu x}$$

For lead, with rays of energy 1·27 MeV, $\mu \doteq 100 \text{ m}^{-1}$

$$\frac{I}{I_0} = e^{-100x}$$

For $x = 0.001$ m, $I/I_0 = 0.90$

$x = 0.005$ m, $I/I_0 = 0.61$

6. This example is very similar to example 15.1.

The energy of the β- particles from the P^{32} is 1.71 MeV, and their RBE $= 1$. The number of disintegrations per second is 3.7×10^6 dps. If the entire energy is absorbed by your body (mass 50 mg) then dosage

$$= 3.7 \times 10^6 \frac{\text{disintegrations}}{\text{second}} \times \frac{60 \text{ s}}{\text{minute}} \times \frac{1.71 \text{ MeV}}{\text{disintegration}} \times \frac{\text{g rad}}{6 \times 10^7 \text{ MeV}} \times \frac{1}{50 \times 10^3 \text{g}}$$

$$= 1.27 \times 10^{-4} \text{ rad/min}$$

Since the RBE factor is 1, this gives a dosage of 1.27×10^{-4} rem/min

$$= 127 \ \mu\text{rem/minute}$$

If all the activity is absorbed by 100 g of intestine, then the dosage rate is 127×500 $= 6.3 \times 10^4$ μrem/min = 63 mrem/minute.

Taking your maximal permissible dose as 8 rem in any 13 week period, you will reach this is $8/63 \times 10^{-3}$ min $= 126$ minutes, or just over two hours. Bad news indeed.

7. The fossil contains 9g of carbon, and thus contains

$$\frac{6 \times 10^{23}}{12} \times 9 \text{ molecules} = 4.5 \times 10^{23} \text{ molecules.}$$

If $10^{-8}\%$ of these were originally C^{14}, then the initial number of C^{14} molecules was 4.5×10^{13} molecules.

The disintegration constant of C^{14} is $0.693/5770$ y

$$= \frac{0.693}{5770 \text{ y}} \times \frac{1 \text{ y}}{365 \text{ d}} \times \frac{1 \text{ d}}{24 \times 3600 \text{ s}} = 3.8 \times 10^{-12} \text{ s}^{-1}.$$

The number of radioactive carbon atoms present now is given by $dN/dt = -\lambda N$

i.e. $N = dN/dt/\lambda = 36/3.8 \times 10^{-2} = 9.4 \times 10^{12}$

The age of the fossil is thus found from $N_t = N_0 \ e^{-\lambda t}$

$$\text{Age} = \frac{5770}{0.693} \ln(4.5 \times 10^{13}/9.4 \times 10^{12}) \doteq 13\ 000 \text{ years}$$

Further Reading

General physics

Physics for Biology and pre-medical students. Burns, D.M. & MacDonald, S.G.G. Addison-Wesley (1970).
Physical Science for Biologists. Edgington, J.A. & Sherman, H.J. Hutchinson (1971).
Lectures on Physics. Feynman, R.P. *et al.* Addison-Wesley (1965).
Physics for the Inquiring Mind. E.M. Rogers, O.U.P. (1960).
Mr Tomkins in Wonderland. Gamow, G. If you want to know what modern Physics is all about with bedtime reading, then this is it.

Background mathematics

Mathematics for Biologists. Crowe, A. & Crowe, A. Academic Press (1969).
Basic Mathematics for Biological and Social Sciences. Marriott, F.H.C. Pergamon (1970).
Calculus made easy. Sylvanus P. Thompson.
Biomathematics. C.A.B. Smith. Griffin.
Quick Calculus. Kleppner, D. & Ramsey, N. Wiley.
Standard Mathematical Tables. Selby, S.M. Chemical Rubber Co. (1969).
Tables of Integrals and other Mathematical Data. Dwight, H.B. Macmillan (1961).

Particular aspects of the book

Animal Mechanics. Alexander, R.M. Sidgewick & Jackson. (1968).
Animal Skeletons. Currey, J.D. Arnold (1970).
Introduction to Polymer Science. Treloar, L.R.G. Wykeham Publications (London) (1970).
General Properties of Matter. Newman, F.H. & Searle, V.H.L. Arnold (1959).
Energy Exchange in the Biosphere. Gates, D.M. Harper & Row, (1962).

Introduction to Thermodynamics. Spanner, D.C. Academic Press (1964).

Agricultural Physics. Rose, C.W. Pergamon (1966).

Biological Effects of Radiation. Coggle, J.E. Wykeham Publications (London) (1971).

Molecular Biophysics. Setlow & Pollard. Addison-Wesley, (1962).

Animal Locomotion. Gray, Sir James, Weidenfeld & Nicholson, (1968).

The Flight of Birds. Brown, R.H.J., Biol. Rev. (1963) 38, 460.

Insect Flight. Pringle, J.W.S., C.U.P., (1957).

Animal Flight. Pennycuick, C.J. Arnold, (1972).

Shape and Flow. Shapiro, A.H. Heinemann, (1961).

Electronics for Biologists. Offner, F.F. McGraw-Hill.

Electronics in the Life Sciences. Young, S.

Electronic apparatus for Biological Research. Donaldson, P.E.K.

Spectroscopy. Whiffen, D.H. Longmans, (1966).

Introduction to electron microscopy. Wischnitzer, S. Pergamon, (1962).

Index